# Quality of Work Life in Libraries

## About the Authors

**Dr. A. Naseer** did his LLB, MLISc and PDEC from the University of Kerala, Thiruvananthapuram. He qualified National Eligibility Test of UGC. He secured a Ph.D. degree in Library and Information Science from the University of Kerala, Thiruvananthapuram. He is a library and information professional with Eighteen years of rich experience in the domain. He trained on Integrated Library Management Software (ILMS), Institutional Repository and e-resources. He has authored several research papers in reputed journals and presented many papers in seminars and conferences. He authored a book. He is the Best Paper Award recipient at the International Conference on CTTL-2019 held in Manonmaniam Sundaranar University, Tirunelveli. His areas of interest are management of library and information centres, human resources development, information and reference management services, and intellectual property. He has also participated in several workshops and training programmes. He is a life member of various professional library associations. He is at present serving as Assistant Librarian at the Kerala Agricultural University, College of Agriculture, Vellayani, Thiruvananthapuram, Kerala

**Dr. B. Mini Devi** (Bhavani Mini Devi) is currently working as Assistant Professor and Former Head of the Department of Library and Information Science at the University of Kerala, India. She is presently the Director of the Centre for Information Literacy Studies at the University of Kerala. Devi is the sole permanent female teaching faculty since 2011. The University of Kerala bagged NAAC A++ Accreditation during her departmental headship. She also chaired as the Director of UIT (University Institute of Technology) of the University of Kerala. She serves as the Chairperson, Board of Studies in Library and Information Science, University of Kerala and member in University of Calicut and Kannur University. She is also a member of the Board of Examiners at the University of Kerala and Calicut University. She widely contributes to the University of Kerala through policy-making, regulation framing, and syllabus updation as the Chairperson and Academic Council Member. She actively participated in sensitisation and grievance redressal as an Anti-Ragging Committee Member of the University of Kerala. She essays the role of Resource Person for the Refresher Courses at the University of Kerala and University of Calicut. She also conducts innovative Conferences, (Online/Offline), Talks and Exhibitions on Library Science. She chairs as the subject rapporteur at various national and regional conferences. She also contributes as a career counsellor, subject expert, library mentor and motivational speaker at various universities and colleges. She scored the third rank in MSc. Botany, NET-JRF along with third rank in M.L.I.Sc. from the University of Kerala. She began her career with the prestigious British Library apprenticeship, Ernakulam Model Engineering College librarianship and 15 years professional experience as librarian. She completed her PhD in "Toxicology Literature: An Informetric Analysis" from Cochin University of Science and Technology, Ernakulam, Kerala, India. She specializes in library classification, cataloguing, informetrics, scientometrics, research methodology, statistics & data analysis, information literacy, media literacy, copyright & intellectual property rights, library management, KOHA, data mining, machine learning, Google SEO, AI, ChatGPT, and information storage & retrieval. She published a wide range of books and book chapters. Alongside 200 co-authored articles, she published over 150 articles in national & international journals and conference proceedings. She produced 20+ doctoral PhD thesis, 2 postdoctoral theses and 5 MPhil dissertations under her research guidance and mentorship. She believes in unbiased, student-friendly teaching through the humble educational motto of the University of Kerala—Karmani Vyajyate Prajna (Wisdom manifests into action).

# Quality of Work Life in Libraries Issues and Prospects

**A. Naseer**
Assistant Librarian
Kerala Agricultural University
College of Agriculture, Vellayani
Thiruvananthapuram

**B. Mini Devi**
Assistant Professor
Department of Library & Information Science and
Honorary Director, Centre for Information Literacy Studies
University of Kerala, Thiruvananthapuram

PMW

PMW
FU 75, Pitam Pura
New Delhi 110034
Mob. 91-8447075807
Email: publishmywork23@gmail.com
Web: www.publish-mywork.com/



Print ISBN: 978-93-61341-39-7

ebook ISBN: 978-93-61344-64-0

# Preface

Quality is a multi-dimensional concept in the context of university education. The functioning of libraries in the universities manifests on such dimensions. The library supports entire academic activities at the university. The library has a vital role in the teaching, learning, research and extension activities as a learning resource centre in the current digital learning environment. Evaluation of libraries is monitored during the accreditation process of academic institutions. Hence, libraries have a significant role in academic and research pursuits both qualitatively and quantitatively.

Presently, the role and functions of libraries in academic institutions have been redefined. Libraries are adopting managerial techniques for enhancing efficiency in operations and services. The success of the libraries nowadays mainly depends on the most effective utilization of its resources and strategic management in the functioning of the libraries. Studies on Quality of Work Life (QWL) in India is minimal about libraries. Hence an attempt is made to investigate the Quality of Work Life in the libraries attached to the universities of Kerala. The study confines the existing state of QWL by measuring it through a modified eight-point criterion advocated by Richard E. Walton. Furthermore, the perception of LIS professionals on job enrichment, job enlargement and users and services have been obtained for conducting this study.

The book has been organized into seven chapters, followed by a Bibliography.

The work is based on research on the area at the University of Kerala. The book is relevant to the current trend of university libraries, especially in adopting management techniques to humanize the workplace and improve the effectiveness of the services. The book is valuable for decision-makers, working LIS professionals at all levels, faculty members, students, and research scholars of LIS and all others involved in the management of libraries and information centres to manage modern managerial methods.

The authors would like to acknowledge the University of Kerala, Thiruvananthapuram, for the facilities and support to carry out this study and bring out this book. The authors would also like to record our sincere gratitude to all the Library and Information Science Professionals working in the universities of Kerala across the state for proving necessary data for this study's conduct. We also express our sincere gratitude to our professional colleagues, friends and well-wishers who have immensely encouraged and assisted us with the publication of this book.

We also take this opportunity to express our special thanks to Publish My Work, New Delhi for bringing out this book in the most professional way.

Thiruvananthapuram **Authors**

# Contents

# List of Abbreviations

| | | |
|---|---|---|
| AICTE | : | All India Council of Technical Education |
| ANOVA | : | Analysis of Variance |
| BLISc | : | Batchelor of Library and Information Science |
| CABI | : | Centre for Agriculture and Biosciences International |
| CAS | : | Current Awareness Service |
| CD ROM | : | Compact Disc Read Only Memory |
| CDC | : | Curriculum Development Committee |
| CeRA | : | Consortium of e-Resources in Agriculture |
| CSIR | : | Council of Scientific Industrial Research |
| CUSAT | : | Cochin University of Science and Technology |
| CUSATL | : | Cochin University of Science and Technology Library |
| DDC | : | Dewey Decimal Classification |
| DDS | : | Document Delivery Service |
| DELNET | : | Developing Library Network |
| ETD | : | Electronic Theses and Dissertations |
| HECI | : | Higher Education Commission of India |
| ICQWL | : | International Council for Quality of Working Life |
| ICSSR | : | Indian Council of Social Science Research |
| ICT | : | Information and Communication Technology |
| IDEAL | : | Indian Digital Ensemble of Agricultural Libraries |
| ILL | : | Inter Library Loan |
| ILMS | : | Integrated Library Management Software |
| INFED | : | Indian Access Management Federation |
| INFLIBNET | : | Information Library Network |

| | | |
|---|---|---|
| IR | : | Institutional Repository |
| IT | : | Information Technology |
| JCCC | : | J-Gate Custom Content for Consortium |
| KAU | : | Kerala Agricultural University |
| KAUL | : | Kannur University Library |
| KAULIS | : | Kerala Agricultural University Library and Information System |
| KU | : | Kannur University |
| KUL | : | Kerala University Library |
| KVASU | : | Kerala Veterinary and Animal Sciences University |
| KVASUL | : | Kerala Veterinary and Animal Science University Library |
| KVASULIS | : | Kerala Veterinary and Animal Science Library and Information System |
| KVK | : | Krishi Vigyan Kendra |
| LIS | : | Library and Information Science |
| LISA | : | Library and Information Science Abstract |
| M.Lib.I.Sc | : | Master of Library and Information Science |
| M.Phil. | : | Master of Philosophy |
| MANOVA | : | Multivariate analysis of variance |
| MGUK | : | Mahatma Gandhi University, Kerala |
| MGUL | : | Mahatma Gandhi University, Kerala Library |
| MHRD | : | Ministry of Human Resources Development |
| MLISc | : | Master of Library and Information Science |
| NARES | : | National Agricultural Research and Education System |
| NEP | : | National Education Policy |
| NET | : | National Eligibility Test |
| NIRF | : | National Institute of Ranking Framework |
| OPAC | : | Online Public Access Catalogue |

| | | |
|---|---|---|
| Ph.D. | : | Doctor of Philosophy |
| QWL | : | Quality of Work Life |
| RARS | : | Regional Agricultural Research Stations |
| SCAIRIRJ | : | Saraswatham Centre for Advanced Information Resources International Research Journal |
| SDI | : | Selective Dissemination of Information |
| SSUS | : | Sree Sankaracharya University of Sanskrit |
| SSUSL | : | Sree Sankaracharya University of Sanskrit Library |
| UGC | : | University Grants Commission |
| UoC | : | University of Calicut |
| UoCL | : | University of Calicut Library |
| UoK | : | University of Kerala |
| VC | : | Vice Chancellor |

# 1

# Introduction

## Introduction

Human beings are the most critical components of an organization. The organization's survival has improved all the facets contributing to its success. In a competitive world, human resources are sine qua non in an organization. Quality of work life (QWL) is critical for deciding organizational employee-related outcomes. Human resource management has a unique role in any organization. Organizational goals in an organization are achieved through planning, organization, staffing, direction, coordination, and control. The success of organizations depends on how they attract, motivate, and retain their personnel. Workplace relations have a significant influence on employees' quality of life. Organizations are creating a platform by setting goals to showcase the employees' skills. Organizational effectiveness and well-being of the employees are prominent characteristics of QWL.

Quality of Work Life (QWL) is a modern management technique for managing human resources in organizations, including service-oriented institutions like libraries and information centres. University libraries play a pivotal role as a service organization, enabling the user fraternity to learn better as they progress academically and acquire information skills. University Libraries are managed by skilled, talented, and professionally competent university personnel. Library and Information Science (LIS) professionals collect, organize, and preserve resources and facilitate the resources to the clientele through reference, instruction, orientation, sensitization, and developing outreach programs to disseminate collections and services of libraries. Exploring and enriching QWL in Libraries and Information centers are retaining competent professionals. Library and Information Science LIS professionals expect professional advancement, and they are also quick learners, great team players, and have a promising role. The fulfillment of the objectives would make the university libraries stronger. It necessitated understanding the Quality of Work Life (QWL) of LIS professionals working in the universities of Kerala.

## Human Resource Management

Human resources are a vital element for the development of any organization. It is designed to maximize the performance of the employees. Each employee has a unique role in the organization. Organizations are set apart for more funds for recruitment, training, and improving personnel efficiency. Organizations are also taking steps to improve the Quality of Work Life (QWL) of employees. It has a direct and positive impact on the employees and the overall development of the organization. It encourages and motivates employees, compels them to remain in the organization, and enables them to balance their personal and work lives. It also allows the employees to work effectively and encourage activities to manage and develop libraries. Libraries are following several innovation projects and human resource management practices to enrich their services.

## Quality of Work Life (QWL): Meaning and Definition

Quality of Work Life (QWL) is an umbrella term. It is an abstract and broad concept. It is significant in many respects. QWL has been defined in many ways. The term QWL emerged during the 1970s. Walton (1975) introduced this concept as an eight-dimensional model. It measures humanistic values and social responsibilities in an organization. Davis (1983) defined QWL as "the favourableness or unfavourableness of a job environment for people". Johannsen and Page (1980) described QWL as the approach to job and organization design seeking to improve the working environment and employer-employee relations. According to Harrison (1985), "Quality of Work Life (QWL) is the degree to which the work of an organization contributes to material and psychological well-being of its members". May et al. (1999) defined QWL as the "favorable conditions and environments of a workplace that support and promote employee satisfaction by providing workers with rewards, job security, and growth opportunities."

In the present study, QWL refers to the totality of feelings and experiences in the life of the selected Library and Information Science (LIS) professionals at work on the job-related factors in university.

## University Library

Harrod's Librarian's Glossary defined University Library as "a library or group of libraries established, maintained, and administered by a university to meet the needs of its students and members of the academic staff" (Prytherch,1995).

In the present study, university libraries refer to the libraries that are functioning under the state universities in Kerala which have been accredited by national-level bodies like the Indian Council of Agricultural Research/University Grants Commission.

## Quality of Work-life in Libraries

Universities are placing emphasis on the quality of research and quality of education. The library is part and parcel of higher education. It is considered the heart of the university. The library is a trinity of resources, staff, and users. Resources of the libraries include books, magazines, media, music, journals, e-books, and databases. Online digital resources are extensively used for teaching, learning, research, and extension activities. LIS professionals in the universities are working with diverse user communities. Libraries are undergoing massive changes across the world. Many changes are initiated in libraries as a result of digital disruption. Technology-enabled information retrieval systems facilitate access, navigation, and retrieval of information. Libraries are hierarchically structured organizations. Human resources are an essential area in Libraries and Information Centres. Librarianship needs specialized knowledge and skills to support and promote the use of library and information systems. The role of professional librarians is unique in a library. A significant portion of the library budget is spent on the salary and allowances of the staff. The employees carry out the work assigned to them based on their skills and professional qualifications. It enabled the fulfillment of the objectives of the libraries and information centers. Therefore, the library has to encourage users' participation and obtain feedback from users for developing and maintaining library services with information and ideas flowing from both libraries to users and vice versa.

University libraries are one of the most enduring support systems of higher education. A qualified LIS professional is an essential component of a university library. The role of librarian has changed from curator to custodian to facilitator. Collection development of the library includes evaluation of user needs and analysis of the weakness of the collection. Preserving library materials includes conserving, repairing, and maintaining the physical and digital collections. LIS professionals create online catalogs, describe material with metadata to enable easy discovery, maintain access to licensed electronic resources, develop systems to organize and present digital collections and establish techniques to discover and harvest digital information. Data management systems of the library collect, describe, and manage data to facilitate user fraternity. The use of social media, open access, and digitization is also enhanced in libraries. In libraries, LIS professionals are categorized as professional, semiprofessional, nonprofessional, and supporting staff in the library. It was further classified as the top, middle, and lower levels. All the staff are responsible at different levels of the hierarchy. The hierarchy position is called the level of management. The level of management in a library depends on the organization's objectives, the size of the collection, the service, and the scope of the work performed by the staff. The library professional staff working at the top level of the university library is called a Librarian. The LIS professionals working in the universities are working hard to enhance the knowledge of the library's clientele. It is the most efficient way to keep up with them in this era of ever-expanding knowledge.

**Statement of the Problem**

The productivity of LIS professionals in university libraries depends on many factors. QWL is found to affect the work of LIS professionals. The workforce is a significant input for the survival and growth of an organization. Dr. S. R. Ranganathan, as early as 1928, rightly pointed out that the 'Library is a growing organism.' He stressed the importance of libraries and their services in the famous book 'Five Laws of Library Science.' 'Five Laws of Library Science' appeared in a book format in 1931. The principles behind the five laws of library science remain unchanged even in the present context of libraries. It reveals that the environment of libraries is changing with the application of advanced technologies and innovation. The law expanded the role of libraries and redefined functions and their services. The fifth law of library science, "Library is a growing organism," is significant in improving the quality of the services rendered by the library. If the working conditions of the LIS professionals are improved, it is possible to fulfill the libraries' vision, mission, and mandate.

There has been a rapid rise in higher educational institutions in the last couple of decades. Nowadays, India has one of the largest education systems in the world. It has brought revolutionary changes to the university education sector. The number of universities and enrollment ratio have been increasing. There is phenomenal growth in the universities regarding infrastructure, faculty, and technical staff. The development of universities can be possible through a well-equipped library. LIS professionals have to play a crucial role in the university setup. It can be achieved through a congenial QWL in university libraries. The authorities of the universities should create an atmosphere to enhance the quality of working life in the libraries of the universities. Many service sectors have improved their services and human resources by implementing quality work-life programmes. Therefore, the present study attempts to probe into all these aspects among the LIS professionals working in the libraries attached to the universities in Kerala. It helps to understand the overall QWL in the libraries.

## Need of the present study

Recently, there have been several reasons to justify the study of Quality of Work Life (QWL) in the university libraries of Kerala. Modern libraries in the university can survive in the higher education sector if they contain only advanced management techniques in their functioning. It is necessary to introduce innovative services and best practices in the libraries of the universities. Digital India is the new slogan. Literacy is the most critical indicator of human capital. Kerala ranks first in the country with a literacy rate of 93.91 percent (2011 Census). The outbreak of the pandemic Covid 19 forced the universities to raise their activities to a different level. It posed many challenges and opportunities to LIS Professionals. It necessitated strengthening the human resources of the library and the performance of the LIS professionals to the occasion. Libraries are sharing and disseminating their resources and services, supporting their technological knowledge and infrastructure. Besides books, journals, and periodicals, university libraries offer various information services manually and electronically. It provides the clientele with valuable information and resources to strengthen their academic activities. In any organization, employees accomplish the objectives of the organization. University libraries have realized the importance of quality of work in Libraries. The libraries are making significant efforts to provide adequate services to the clientele. The need to diagnose the existing QWL in university libraries necessitated improving the working conditions in the libraries. It has necessitated a study on the Quality of Work in Life in the University libraries of Kerala.

## Objectives of the study

The study's main objective is to investigate the QWL of the (Library and Information Science) LIS professionals working in the university libraries of Kerala, India. In the present study, the following objectives are specifically formulated from the main objective.

1. To understand the Quality of Work Life (QWL) in the university libraries of Kerala.
2. To study the Quality of Work Life (QWL) among LIS professionals working in the University Libraries in Kerala.
3. To compare the situation of QWL in the university libraries of Kerala.
4. To study the extent of job enrichment in the university libraries of Kerala.
5. To study the extent of job enlargement in the university libraries of Kerala.

6. To investigate the perception of the library and information professionals towards the users and services in the selected university libraries.
7. To suggest measures for strengthening Quality of Work Life (QWL) in the libraries of Universities of Kerala.

## Hypotheses of the study

The following hypotheses are formulated for the study in light of the above objectives.

H1: Quality of Work Life (QWL) in the libraries attached to the universities are the same.

H2: There is a significant association in QWL among LIS professionals in the universities of Kerala.

## Scope of the Study

There are 14 state universities functioning in Kerala (Economic Review 2020). However, the present study is focused on the libraries attached to the eight major state universities in Kerala. The universities selected for the current study are:

| Sl.No. | Name of the University | Year of Establishment |
|---|---|---|
| 1. | University of Kerala, Thiruvananthapuram | 1937 |
| 2. | University of Calicut, Thenjippalam, Malappuram | 1968 |
| 3. | Kerala Agricultural University, Vellanikkara, Thrissur | 1971 |
| 4. | Cochin University of Science and Technology, Kochi | 1972 |
| 5. | Mahatma Gandhi University, Kottayam | 1983 |
| 6. | Sree Sankaracharya University of Sanskrit, Kalady, Ernakulam | 1993 |
| 7. | Kannur University, Kannur | 1998 |
| 8. | Kerala Veterinary and Animal Sciences University, Pookode, Wayanad | 2010 |

As mentioned above, the universities have well-established libraries, a rich collection of resources in different subjects, and the application of advanced technology in their services. The study respondents are the personnel from the cadre of Library Assistants/Professional Assistants Grade II to the cadre of Deputy Librarians/University Librarians in charge working in the university libraries.

## Utility of the Study

The study throws light on the existing state of Quality of Work Life (QWL) prevailing in the libraries of the universities of Kerala. It helps libraries and information centers fulfill the organization's objective and enhance the effectiveness and services of the libraries attached to the universities.

**Size and population of the study**

Fourteen state universities are functioning in Kerala. For the present study, the research was conducted in the eight major state universities in Kerala. The Census method of research was used for the collection of data. There are three hundred and eighty-five (385) sanctioned posts of Library and Information Science (LIS) professionals in all these universities. All the universities selected for the study have their libraries. Quality of Work Life (QWL) was measured with a modified QWL scale based on Walton's eight-point criteria. In the present study, there were only two hundred and seventy-six (276) LIS working in the selected universities. The remaining post of LIS professionals was lying vacant. For the detailed study, a structured questionnaire was prepared. The questionnaires were distributed to all the 276 LIS professionals working in the universities. Out of the 276 respondents considered for the study, only two hundred and sixty-six (266) respondents duly filled and returned the questionnaire. The response rate was 96.38 percent. Data obtained from all 266 LIS professionals are finally selected for conducting the present study. Thus, the total population available for the present study is 266 LIS professionals working in the eight major state universities of Kerala. The responses obtained from the questionnaire were based on the analysis and findings of the present study.

**Methodology**

The data for the present study was collected through the following methods.

**Review of Literature**

Literature on the related topic has been reviewed for the preparation of a tool for the study. The literature search was done among primary, secondary, and tertiary sources of information.

**Tool Used**

The questionnaire was used to elicit information on the various aspects related to QWL from the LIS professionals working in the universities of Kerala. Modified Walton's eight-point dimension was used to collect information about QWL.

### Data Analysis

Data collected through questionnaires were tabulated, analyzed, and interpreted using the latest version of Microsoft Excel and Statistical Package for Social Science. (SPSS, Version,23). Appropriate mathematical and statistical techniques are also used.

### Study Period

The period of the present study spans from 2014 to 2019. The process of collecting data for the present study commenced in September 2018 and ended in March 2019. However, the data collected for the present study was mainly from September 2018 to March 2019.

### Limitations of the Study

1. The study has been limited to the libraries attached to the eight state universities in Kerala.
2. Only regular LIS professionals working in the universities are selected for the study.
3. The LIS professionals appointed on a daily wage and contract basis are excluded from the study.
4. The present study covers the status of a particular point in time.

### Organization of the Study

The entire theme of the present study is analyzed, presented, and structured into the following seven chapters, followed by a bibliography. It covers various aspects of Quality of Work Life (QWL) among LIS professionals working in university libraries.

Chapters – 1 Introduction

The chapter introduces all the critical areas under the topic that have been mentioned to get an understanding of the quality of work-life. It contains an introductory part, Quality of Work Life (QWL) among the employees working in the university libraries, objectives of the study, hypotheses of the study, scope and limitations of the study, and other general information relating to the topic that has been mentioned.

Chapter – 2 Review of Literature

Chapter 2 deals with the review of literature and studies of different related aspects of the subject. Studies on the research topic at the national level and international levels were collected and analyzed to get an overview of the study.

Chapter – 3: Historical Development of University Libraries in India

The chapter highlights the historical development of libraries, committees, and commissions on university libraries in India.

Chapter – 4 Quality of Work Life: An Overview

The chapter deals with the quality of work life presented as a detailed subject.

Chapter – 5 Profiles of Selected University Libraries in Kerala

The chapter accounts for the relevant information related to the profiles of selected eight state university libraries in Kerala.

Chapter – 6 Data Analysis and Interpretation

The chapter is devoted to data analysis and interpretation relating to the study. The data collected through the questionnaire was tabulated and exposed to specific statistical techniques, inferences were derived, and conclusions were made.

Chapter – 7 Findings and Conclusion

The chapter presents the study's significant findings under each objective formulated for the study, testing of hypotheses composed, research findings, suggestions, further areas of research in the topic to be chosen, and conclusion. The information in this chapter is derived from the entire work.

## References


Davis, L. E. (1983). Design of new organizations. In Harvey, K., Hansvan, B. (eds.). The Quality of Working Life and the 1980s. Praeger Publishers, 65–86.

Economic Review,2020(2021). Thiruvananthapuram: Kerala State Planning Board. Vol. I.

Harrison, T.M., (1985). Communication and Participative Decision Making: An exploratory study. Personnel Psychology. (1), 97-116.

Johannsen, H. and Page, G. T. (1980). International Dictionary of Management. New Delhi: Vision Books; p. 277. https://doi.org/10.1007/978-1-349-05769-6

May, B., Lau, R., & Johnson, S. K. (1999). A longitudinal study of quality of work life and business performance. *South Dakota Business Review, 58,* 1-7.

Neelameghan, A. (2007). Library is a growing organism – S R Ranganathan. DLIBCOM, 2 (8-9), 2-3.

Prytherch. Ray. (19950. University Library. In Harrod's Librarians' Glossary (8th ed., p.663). England: Gower.

Ranganathan S R. (1957). The Five Laws of Library Science, 2nd ed. Bombay: Asia Publishing House.

Walton, R.E. (1975). Criteria for Quality of Working Life. In L.E. Davis, A.B. Cherns and Associates (Eds.), The Quality of Working, New York: The Free Press, Life, Vol. 1, pp. 91- 104

# 2

# Review of Literature

## Introduction

This chapter reviews literature related to the current study on Quality of Work Life (QWL). Scrutiny of the available literature is of great value in gaining insight into the research problem. It is beneficial for strengthening the research problem. As there are fewer studies on QWL in the university libraries in Kerala, the works reported in other places were also reviewed. Hence, the available studies related to the topic are exhaustively reviewed. The review covers studies on the subject made at the regional, national, and international levels.

To survey the previous studies on the topic, a number of primary, secondary and tertiary sources were consulted. The information sources include books, postgraduate and doctoral studies, seminar papers, commission reports, articles in primary periodicals, secondary sources like Library and Information Science Abstract (LISA), ProQuest, Information Science Abstract, LISA plus, abstracts and contents of full-text papers available through J gate plus, j store, EBSCO, Emerald, and other web resources which are relevant for the study.

The reviews are arranged under the following headings

- Quality of Work Life in General
- Quality of Work Life in Libraries

## Quality of Work Life in General

Ghasemi et al. (2021) evaluated Quality of Work Life (QWL) among surgeons and investigated its association with musculoskeletal complaints. This study was conducted among surgeons. QWL was assessed using Walton's 35-item questionnaire. In this study, seventy-four surgeons have participated. Most of them had a moderate level of QWL. The development of human capacities and the total life space had the highest and lowest scores. The study finds that QWL significantly predicts pain and discomfort complaints in the neck, shoulders, upper back, elbows, and legs. Improving workstations and tools used by surgeons and interventions aimed at promoting QWL may be necessary to reduce musculoskeletal complaints.

Fakhri et al. (2020) investigated the Quality of Work Life (QWL) among employees of consulting service management in Indonesia. The factors of QWL consist of nine components. They are work environment, organizational culture and climate, relation and cooperation, training and development, reward and compensation, facilities, job satisfaction and job security, work autonomy, and resource adequacy. The study includes gender, education level, employment length, and the QWL condition. The results showed that most dimensions of QWL are within a valid category and indicate that the major factors constituting the QWL are supportive organizational culture and organizational facilities.

Bhende et al. (2020). unearthed dimensions of quality of work-life and work–life balance and found the impact of the quality of work-life on work–life balance. Data have been gathered from 89 managers of public and private sector banks in India. The study indicated that the productivity dimension of a work–life balance was influenced by all dimensions of quality of work-life except grievance redress. Further, all three dimensions of quality of work life predicted the skill deployment dimension. However, none of the quality of work-life dimensions had any relation with the efficiency dimension of work–life balance.

Tamsah et al. (2020) conducted a study to reveal the relationship between training and employee performance in the public sector, with knowledge-sharing indicators as a mediator for both. The sample was selected using a purposive sampling technique to the Civil Servants (CS) questionnaire. The results showed that as many as 149 samples answered the questionnaire and had attended leadership training and other training. The results showed that the role of leadership training and additional training for training assessment tools to increase knowledge sharing was very influential compared to functional training, technical training, and other training. Although leadership training and further training significantly impact increasing knowledge sharing, they have not improved work-life and CS Performance quality. On the other hand, knowledge sharing indirectly affects CS performance compared to its direct impact.

Agrawal, A (2019) investigated the Quality of Work Life (QWL) of employees and its impact on organizational performance. The research was carried out with the cross-sectional data collected from one hundred and fifteen employees from a public sector undertaking. The data was measured with a six-dimensional scale. The data was further validated with CFA, and the structural models were tested using SEM. The individual processes were further analyzed with correlation analysis. The results obtained are unique. It can be used for policy formulations in public sector undertakings.

Suresh, K (2019) investigated the Quality of Work Life among nurses in Private Hospitals. Data were collected from 168 nurses from various private hospitals using a structured questionnaire. The data were analysed using appropriate statistical tools. The study finds that remuneration, opportunity for personal growth, work environment, support leadership and structure, and work-life balance influence the Quality of Work Life. The research provides insights into solving issues relating to Quality of Work Life.

Darmawansyah et al. (2018) analyzed the effect of work stress, quality of work life, and organizational climate factors, which mostly affected officials' work satisfaction at the regional public hospital of Undata Palu. The sample consisted of 247 officials, and they were selected using the sampling method. The data was obtained using a questionnaire. The result indicates that work stress, quality of work life, and organizational climate simultaneously significantly affect officials' work satisfaction. The dominant effect of job stress, quality of work life, and organizational climate on officials' work satisfaction is job demand, balanced compensation, and recognition, respectively.

Ramawickrama; Opatha, and Pushpakumari (2018) investigated the relationship between the dimensions of quality of work life and job performance dimensions to 280 Station Masters at Sri Lanka Railways. The quality of work-life was measured by using Walton's model. A three-component model, including task performance, citizenship performance, and counterproductive work dimensions, was used for measuring job performance. The result of the study revealed that the quality of work life substantially influences job performance.

Shanmugam and Bharathi (2017) conducted a study on the Quality of Work Life (QWL) among self-financing engineering college teachers. The study consisted of all the teachers working in 28 engineering colleges. Simple random sampling was adopted to derive 146 samples. A structured questionnaire was used to collect data. The study finds significant differences in Quality of Work Life. Perceptions were observed, and it was also found that the majority of the teachers belong to a distinct generation. It also revealed that there is a moderate level of satisfaction among teachers on Quality of Work life.

Seema and Maryam (2013) explored various dimensions of the Quality of Work Life as it affects the life and attitude of teachers at private universities at work. A survey was conducted with 360 faculty members to determine their perceptions of QWL and its spill-over effect on employee commitment, engagement, job involvement, and reputation. The scholarly contribution highlighted that the dominant construct of QWL plays an important role in shaping attitudes towards work, life, and teachers' relationships. The study hints at possible implications of dissatisfaction and imbalance within employee commitment and engagement and even the university's reputation on a practical level.

Netto (2017) analyzed the quality of work life of women employees in various private sector organizations. The study was conducted among the ninety-six women employees. The study was analyzed based on demographic factors. Walton's eight-dimension scale was utilized to measure Quality of Work Life (QWL) perception. The study revealed that a significant difference existed between marital status for work and total life space, educational qualification for social relevance on work life, organization type for adequate and fair compensation, constitutionalism, and social relevance of work life, work experience for social integration in the work organization; and monthly income for adequate and fair compensation, opportunities to use and develop human capacities, opportunity for continued growth and security, and social relevance of work life. There was a significant positive relationship between the overall QWL and the majority of each dimension of QWL.

Anandh and Gunasekaran (2016) examined the impact of QWL on the Work Life Balance (WLB) among civil engineering professionals from the construction industry in Chennai. The research followed a sample size of 500 professional engineers. The questionnaire was used to collect the primary data. The study's outcome is that there has a high effect of QWL on WLB. It implies that the satisfaction of employees may balance their personal and professional life in a better way.

Varadaraj and Parthiban (2009) undertook a study to identify the impact of demographic variables on the perceived Quality of Work Life (QWL) of supervisors working in the engineering industry in the Coimbatore District. The study indicated that only gender and age significantly contribute to the QWL of supervisors. The study may help employers to redesign their policies towards the supervisor.

Jahani, M. A., Ghahramanmahmoudi, Yaminfirooz, M., and Shahrdami, Y. (2017) studied the relationship between perceived organisational support and the quality of working life of 350 employees in the five selected hospitals of the north of Iran. A questionnaire was used to collect data. The final model and the model verification were evaluated based on confirmatory factor analysis. Perceived organisational support has been found to relate to the quality of working life and its related dimensions. The study found that perceived organisational support is correlated with working life. It is concluded that the increase in perceived organisational support of staff could improve the quality of their working life.

Ahmadian., Vafaeian., and Farshbaf (2015) determined the relationship between quality of working life with organizational commitment and employee productivity of Physical Education faculties in the Islam Azad Universities of Tehran. The method of research is a correlation. The community includes all the employees in the universities. The sample size of the population was 267 individuals. A questionnaire was used to collect data. The study reveals a positive and significant correlation between the quality of working life and organizational commitment and employee productivity.

Kamboj., Ramnath, and Singh (2016) explored the relationship between emotional competence and the quality of work life of call centre employees from the different cities of Haryana. The study was based on a sample of 150 employees from 4 call centres based on stratified random sampling. The study finds a positive correlation between emotional competence and different aspects of quality of work life.

Nanjundeswara Swamy, T S & Swamy, D R. (2013) conducted a study to know the level of perception towards Quality of Work Life (QWL) of technical institution employees and enhance the management. The sample consists of 109 employees. The questionnaire was based on nine components of QWL. The study reveals that male employees are more satisfied than female employees. There is no significant relationship between the demographic characteristics of employees and QWL. Research also indicates that adequacy of resources are more correlated, and training and development are less correlated with QWL in teaching staff, and in the case of non-teaching staff, compensation and rewards are more correlated, and work environments are less correlated with QWL. The study reveals that all the dimensions of QWL positively correlate with the QWL of faculties. It indicates that enhancement in the dimensions of QWL can increase the overall QWL of faculties.

### Quality of Work Life and Libraries

Bishop., Orehek and Collier (2021) conducted a study to assess the job tasks of earth science data managers and librarians. The study was done through interviews with six librarians and eight data managers. The study found that most of the knowledge, skills, and abilities of workers were acquired through on-the-job experience. However, future professionals may benefit from tailored education informed through job analysis.

Naskar., Hasan and Das (2021) highlighted the popularity of emerging trends and technologies in the Library and Information Services course. The study investigated learner involvement using YouTube Channels and Discussion Forums. The study was analyzed by using social media channels. The study reveals that learners' participation in the online discussion forum increased from time to time, and the social media involvement was also popularized through the YouTube Channel.

Gupta and Kumar (2020) examined students' perceptions of different residential backgrounds for implementing various tools for marketing library resources and services from the eight state university libraries of North-Western India. The research was conducted by administering a questionnaire. The study revealed that libraries must employ Web 2.0 technologies, and most respondents have learned about this form of promotion. It was also found that despite the background, except for a few marketing tools, there remained no major difference in the opinion of users regarding the adoption of various marketing tools.

Omole (2020) investigated the relationship among job satisfaction, leadership styles and career indecision of library personnel. The population consists of 105 library personnel from five selected libraries in private universities in South-West, Nigeria. The questionnaire was used as an instrument. The study finds a significant relationship between job satisfaction and career indecision. Furthermore, there was no significant relationship between leadership style and career indecision. Finally, there was a significant relationship between job satisfaction and leadership style toward predicting career indecision.

Reghagwa and Ijiekhuamhen (2020) assessed the influence of ICT competencies on task performance. The study consists of 235 librarians in the university libraries of South-South, Nigeria. The questionnaire was adopted for the collection of data. The study revealed that the librarians are above average in the use of ICT in libraries. It was also revealed that ICT has improved task performance and reduced the time in executing tasks by creating ample time for engaging in other productive activities. The study also showed other problems that hindered the acquisition of ICT skills by librarians in universities.

Rachel and Mini Devi (2019) measured job satisfaction among library professionals. The study was conducted at the University of Calicut. Job satisfaction is a feeling which a variety of factors could cause. Six components of job satisfaction are derived from the literature. The questionnaire was used to collect the data. The study finds that the library professionals are satisfied with the nature of their work. However, salary, promotion, and denied access to benefits were identified as major constraints to job satisfaction.

Rao (2020) investigated the Quality of Work Life (QWL) that motivates library professionals to learn further and develop skills in the 20 engineering colleges of the Jaipur region, and 65 library professionals were selected. The data has been collected using a structured questionnaire. For this purpose, library professionals have been surveyed. Likert's 5-point scale has been used. A questionnaire was used for data collection, which has been analyzed using statistical techniques. It was observed that library professionals found good job exposure and agreed on satisfactory relationship factors of QWL.

Rao and Verma (2019) measured the Quality of Work Life (QWL) of library professionals to identify the relationship between the QWL of library professionals and factors such as interpersonal relationships, status, service conditions, promotional policies, and motivation for employee's future role and develop skills. Library professionals working in the engineering institutes of the Jaipur region were surveyed using a questionnaire. The study finds that job security and interpersonal relations with senior staff affect QWL.

Koohbanani., Zarei, and Erfani (2019) conducted a study on the Quality of Working Life (QWL) of librarians in Iranian public libraries. The study was conducted among 212 women and 121 men librarians. The Walton standard questionnaire was used as a tool for data collection. The study revealed that dimensions of organizational social integration, constitutionalism in organization, and capacity for human development are at a good level. The dimensions of environmental opportunity for continued growth and security and safe and healthy work are moderate. Finally, dimensions of fair and adequate payment, work life's social relevance, and total life space are at a weak level. The study indicated that librarians have a desirable/high QWL.

Garg and Surya (2018) evaluated organizational management provided by employers and its correlation with the Quality of Work Life (QWL) for Library and Information Science (LIS) professionals. The study analyzed relationships in the LIS professionals' field and determined factors that accentuate their connection. The study collects responses from various LIS professionals and concludes that the organizational environment impacts the establishment of QWL among LIS professionals.

Mothukur and Ramesh (2018) conducted a study on library professionals' Quality of Work Life in ten autonomous engineering college libraries using a questionnaire. The study revealed that most of the respondents have no permanent employment status and that more than fifty percent of the respondents expressed dissatisfaction with salaries on time.

Through a comprehensive literature review, Tiwari and Singh (2018) explored the views or works on the constituent components of human development. The study is divided into seven components. The study includes human resource development policies in various organizations in India and abroad. The components identified in the study are human resource planning, recruitment and selection, placement and induction, and promotion and motivation. The literature indicated that HRD is much more than the employer-employee relationship. The literature reveals that the development of an organization depends on human resources.

Anyaoku (2016) attempted to measure the Quality of Work Life (QWL) of librarians and ascertain the relationship with some demographic characteristics of various library types in Nigeria. The research study used the survey method. The questionnaire, which included a Quality of Life Scale, was used to collect data from 175 working librarians. The study finds that the QWL of librarians is relatively high in areas of opportunity for continued growth and security, social integration in work organization, and social relevance of work. However, important areas such as fair remuneration, training and retraining, provision for job performance, and equal rights issues were dissatisfaction areas for librarians. The study reveals a significant relationship between QWL and gender, age, institution type, and years of work experience. There was no relationship with educational qualification.

Shyni and Haneefa (2016) assessed the role of job satisfaction on the Quality of Work Life (QWL) of library professionals in university libraries. Minnesota Job Satisfaction Questionnaire was used to measure job satisfaction, and the QWL of library professionals was measured with a QWL scale constructed based on Walton's QWL dimensions. Data were collected from 184 library professionals working in the eight university libraries. The study revealed that job satisfaction influences the QWL of library professionals in university libraries. The study also shows that library professionals with different job satisfaction levels have different perceptions of the dimensions of QWL.

Reddy (2010) dealt with the quality of professionals' work-life in Andhra Pradesh public libraries. The study was confined to the professionals working in the State Central Library, Regional Libraries, and Libraries under Zilla Grandhalaya Samsthas. Walton's eight-point criteria, with a few additions, have been taken to measure QWL. This study covered economic, social, psychological, and organizational aspects of work life.

## Conclusion

The preceding literature review aims to review the development of the quality of work life and identify the factors affecting it. The literature enriched knowledge in the subject area of the study. After reviewing the literature available in the field of Quality of Work Life, it is found that studies have been made in other areas and on different aspects of Quality of Work Life in

Libraries. However, no specific study was conducted in the selected university libraries of Kerala. Hence, the present study has been proposed as a pioneering study in university libraries.

## References


Agrawal, Amee (2019). Management Perceptions of the Impact of Quality of Work Life on Organizational Performance. Prabandhan: Indian Journal of Management.12(8),32-45.

Ahmadian, R., Vafaeian, M., & Farshbaf, M. (2015). The Relationship between Quality of Working Life with Organizational Commitment and Employees Productivity of Physical Education Faculties in the Islamic Azad Universities of Tehran. International Journal of Life Sciences, 9(4), 49-53.

Anandh, K. S., & Gunasekaran, K. (2016). Constructing a Model to Examine the Influence of Quality of Work-Life on Work-Life Balance-Discernment of Civil Engineers from Construction Industry in Chennai. Indian Journal of Science and Technology, 9(40)

Anyaoku, Ebele N. (2016) Demographic Determinants of Quality of Work Life of Librarians Working in Nigeria. International Journal of Advanced Library and Information Science, 4(1) :312-323.

Bhende, Pravin et al. (2020). Quality of Work Life and Work–Life Balance. Journal of Human Values. 26(3), 256–265.

Bishop, B. W., Orehek, A. M., & Collier, H. R. (2021). Job Analyses of Earth Science Data Librarians and Data Managers. Bulletin of the American Meteorological Society, 102(4), 1-22.

Darmawansyah, Rompu., Wahyu Atjo., & Abadi M.Y. (2018). The effect of Occupational Stress, Quality of Work life and Organizational Climate on Officials' Work Satisfaction of Regional Public Hospital of Undata Palu, Indian Journal of Public Health Research & Development, 9 (3),309-313doi: 10.5958/0976-5506.2018.00228.0

Fakhri, Mahendra et.al. (2020). Characteristics of Quality of Work Life on Employees at Consultant Company in Indonesia. Journal of Asian Finance, Economics and Business, 7 (11), 1105–1111.

Garg, Babita. & Surya Nath Singh. (2018). Measuring the Relationship between Quality Work Life and Organizational Environment: A Case Study of Library and Information Science Professionals of Haryana State. IOSR Journal of Humanities and Social Science (IOSR-JHSS), 23(10) :14-19.

Ghasemi et al., (2021). Quality of work life among surgeons and its association with musculoskeletal complaints. Cogent Psychology., 8. pp.1-16,8: 1880256

Gupta, R., & Kumar, S. (2020). Students Perception for Adopting Marketing Tools in University Libraries on the basis of their Backgrounds. A Study of North Western India. DESIDOC Journal of Library and Information Technology, 40(6), 339-344.

Jahani, M. A., Ghahramanmahmoudi, Yaminfirooz, M., & Shahrdami, Y. (2017). Structural Model of the Relationship between Perceived Organisational Support and Quality of Working Life of Employees of the Selected Hospital of NorthofIran. Ambient Science, 4(1),23-27.

Kamboj, S., Ramnath, & Singh, S. (2016). Emotional Correlates of Quality of Work Life: A Study on Call Centre Employees. Indian Journal of Health and Wellbeing, 7(9), 940-942.

Koohbanani, Hamta Kazemi. & Zarei, Atefeh. (2019) Measuring the librarians' quality of working life in the Iranian public libraries, Library Management,40(8,9),532-542. © Emerald Publishing Limited, 0143-5124. DOI 10.1108/LM-11-2018-0089

Mothukur, Anjaiah. & Ramesh, Elagandula. (2018). Quality of Work Life of Library Professionals in Selected Autonomous Engineering College Libraries of Hyderabad. International Journal of Library and Information Studies, 8(2): 339-345.

Nanjundeswara swamy, T S & Swamy, D R. (2013). Quality of Work life of Employees in Private Technical Institutions. International Journal for Quality Research 7(3) 3–14.

Naskar, D., Hasan, N., & Das, A. K. (2021). Pattern of Social Media Engagements by the Learners of a Library and Information Science Mooc Course: An Analytical Study. Annals of Library and Information Studies, 68(1), 56-66.

Netto C S (2017). Quality of Work Life of women employees in private sector organizations in Ernakulam District of Kerala. Prabandhan Journal of Management.11(11),51-62.

Omole, Victoria Imabong (2020). Relationship among Job Satisfaction, Leadership Styles and Career Indecision of Library Personnel in Selected Private Universities in South-West, Nigeria. International Journal of Library Science, 9(2): 40-50

Rachel, Rabeka., & B, Mini Devi, B (2019). Job Satisfaction of Library Professionals of University of Calicut: An Analytical Study. Indian Journal of Information Sources and Services, 9(1), 132-135.

Ramawickrama, J; Opatha, H.H.D.N.P & Pushpakumari, M D (2018). Quality of Work Life and Job Performance: Empirical Evidence from Station Masters Working at Sri Lanka Railways. Sri Lankan Journal of Human Resource Management, 8 (1) ,1-21.

Rao, Mahendra Singh (2020). Quality of Work Life (QWL) Motivates Library Professionals Highly to Learn Further and Develop Skills. Library Philosophy and Practice (e-journal). 4618. https://digitalcommons.unl.edu/libphilprac/4618

Rao, Mahendra Singh. & Verma, L. N. (2019). Quality of Work Life among Library Professionals: A Study of Determinant Factors. SRELS Journal of Information Management, 56(2):95-99.

Reddy, Pratapa B. (2010). Quality of work life of professionals in public libraries in Andhra Pradesh - A study. PEARL - A Journal of Library and Information Science,4(2), 95-103.

Reghagwa, I. C., & Ijiekhuamhen, O. P. (2020). Influence of Information Communication Technology Competencies on the Task Performance of Librarians in University Libraries in South- South, Nigeria. Library Progress, 40(2), 204-214.

Seema, Arif., & Maryam, Ilyas. (2013). Quality of work-life model for teachers of private universities in Pakistan. Quality Assurance in Education, 21(3), 282 – 298. http://dx.doi.org/10.1108/QAE-Feb-2012-0006.

Shanmugam, S &. Bharathi, Subburethina P (2017). A Study on the Influence of Socio-Economic Factors on Quality of Work Life Perceptions among Self-Financing Engineering College Teachers in Tiruchirappalli District. Shanlax International Journal of Management,5(1),147-158.

Shyni, K.G. & Muhammed Haneefa, K. (2016). Influence of Job Satisfaction on Quality of Work Life of Library Professionals in University Libraries in Kerala. KELPRO BULLETIN, 20 (2): 64-71.

Suresh, K (2019). Empirical study on the Quality of Work Life of Nurses in Private Hospitals. Shalax International Journal of Management.7(1),92-102.

Tamsah, H et.al (2020). Training, Knowledge Sharing, and Quality of Work-Life on Civil Servants Performance in Indonesia. Journal of Ethnic and Cultural Studies, 7(3), 163-176.

Tiwari, N., & Singh, K. P. (2018). Human Resource Development in Libraries: A Literature Review. Library Herald, 56(2), 241-254.

Varadaraj, S., & Parthiban, V. 2009. Impact of Demographic Variables on Quality of Work Life of Supervisors. SMART Journal of Business Management Studies, 5(2),47-52.

# 3

# Historical Development of University Libraries in India

## Introduction

It has been proved that libraries play a crucial role in education. Libraries have been collecting, preserving, and organizing knowledge since time immemorial. In the age of virtual information, libraries have been reinventing themselves as accessible and engaging learning spaces. Libraries existed in India during the ancient and medieval periods. The library is the primary requirement of the university. University libraries have a paramount role in developing university education in the country. The present chapter highlights the historical development of libraries, committees, and commissions on university libraries in India.

## Genesis and Development of Libraries

Libraries have a long history, from ancient times to present-day high-tech libraries. The development of libraries is related to the development of civilization. Librarianship is as old as libraries themselves. Libraries and librarianship are important parts of the information world. The ancient civilizations of Babylonia, Mesopotamia, Egypt, and Assyria had libraries. India has a great tradition of learning and teaching. Archaeological and literary documents reveal that writing and reading have been practiced since ancient times. Indus Valley civilization flourished in India. Libraries existed during Indus Valley Civilization. The libraries in the Indus Valley civilizations were called "Pustakalayas".

Learning and teaching have been traditions in India since ancient times. "In olden days, teachers of India were themselves Librarians, and they were held in the highest esteem". However, during this period, there were no records of academic libraries because the teachers were named or considered 'memory libraries'. Unfortunately, only privileged people like Brahmans and Kshatriyas could take advantage of formal education. (Kaula:1958)

Viharas and monasteries came into existence during the Buddhist period. Nalanda, Taxila, Banaras, Vikrasmsila, Mithila, Jagadala, Nadia, Ratnagiri, and Lalitri are the famous seats of higher learning. Nalanda and Taxila had their libraries. Nalanda has a library that is situated in a particular area known as Dharmanganja (Mart of Religion). Acharya Nagarjuna, the founder of Mahayana Buddhism, is known to have maintained a library.

All leading Mughal rulers in India are maintained libraries. The Bahamani Kings of Deccan in South India established many academic institutions with libraries in their kingdom. In the Sangam age, there is a considerable tradition of libraries in South India. In the British Period, the colleges established in the 1800s by the Christian missionaries at Calcutta, Agra, Madras, Bombay, etc., maintained good libraries.

## Universities in India

India has the most extensive higher education system in the world. Educational institutions have increased after independence. There is an increase in the number of teachers and enrolled students in higher education.

**Table 3.1:** Total Number of Universities in the Country

| Sl.No. | **Universities** | Total number |
|---|---|---|
| 1. | State Universities | 418 |
| 2. | Deemed to be Universities | 125 |
| 3. | Central Universities | 54 |
| 4. | Private Universities | 370 |
| | Total | 967 |

Table 3.1 demonstrates comprehensive statistical data about the universities in the country as of 31 December 2020. It shows a total 967 number of universities in the country in all four categories. Out of these, 418 universities are state universities,125 institutions are under-deemed to be universities,54 universities are under central universities, and the remaining 370 universities are under private universities.

## Development of University Libraries

The first three universities were started at Calcutta, Bombay and Madras in1857. However, the libraries of the universities are subsequently coming into existence. Calcutta University library is the first university library established during the reign of the British in India. India had only twenty universities during the British period from 1857 to 1947. Almost all the universities during the period had libraries.

Although the University of Calcutta was launched in 1857, the attempt to initiate the library was started in 1872. The Calcutta University Act of 1904 empowered the university to maintain libraries. In 1937, Calcutta University appointed a professionally qualified librarian. Madras University Library was opened to the public in 1907. Madras University library is called the mother library of all the academic libraries of South India. In 1924, Dr S. R. Ranganathan joined the Madras University Library as a Librarian. The University of Mumbai (Bombay) Library was opened to readers in 1880. The Indian University Act (1904) was passed with statutory provisions for universities and affiliated colleges to maintain libraries.

### University Libraries in India in the Post-Independence Period

There has been a phenomenal growth of university education and university libraries in India since independence. The University Grants Commission (UGC) has paid attention to improving libraries. There has been drastic progress in university libraries after independence.

### The University Education Commission (1948-49)

The University Education Commission (1948-49), under the chairmanship of Dr. S. Radhakrishnan, stated that the "Library is the heart of all university's work." The University Education Commission (1948-49) proposed expanding and improving higher education. The committee stressed the need for strengthening academic libraries. The committee recommended that 6.25 percent of the university's total budget, or Rs. 40.00 per student, be allocated as the annual grant for the libraries. The committee also stressed the need for an open access system, well-qualified staff, and twelve working hours a day. The recommendation of the Commission in 1949 was comprehensive, covering all aspects of university education in India (Ghosh,2007)

### Role of UGC in the Development of University Libraries

University Grants Commission (UGC) came into existence in 1953. The UGC became a statutory organization of the Government of India by an act of Parliament in 1956 to determine the standards of universities and the promotion and coordination of university education in India. Dr. C. D. Deshmukh, the first Chairman of UGC, drew an analogy between a university and the human body, remarking that "the students form the body of the university, the administration is the head, the teachers are the soul, and the library is the heart."

UGC has decentralized its operations by setting up six regional centers at Pune, Hyderabad, Kolkata, Bhopal, Guwahati, and Bangalore. The head office of the UGC is located at Bahadur Shah Zafar Marg in New Delhi, with two additional bureaus.

The UGC played a crucial role in the improvement of university libraries. UGC is funding enormously for developing universities and their university libraries in the country. The library-related commissions and committees of the UGC are:

**The Library Committee (1957)**

UGC appointed a commission under the chairmanship of Dr S. R. Ranganathan in 1957 to advise on matters relating to the functioning and management of college libraries and university libraries in the country. Other members of the UGC Committee were Professor S. Bashiruddin, K. S. Hingwe, B. S. Keshavan and S. Parthasarathy. The committee submitted its report in 1959 with several recommendations.

The Committee submitted its report in 1959 covering various aspects of the library. The Library Committee suggested looking into the standards of teaching, examination, and research in library schools. It also includes, a university should appoint full-time library science faculty for teaching, and the librarians should possess prescribed professional qualifications. The Committee observed that the library staff's status and salary should be the same as that of the teaching and research staff. The UGC accepted the recommendation of the Library Committee (1957) as a norm for promoting university libraries.

**UGC Review Committee on Library Science (1961)**

UGC Review Committee on Library Science was appointed in 1961 and submitted its report in 1962. Dr S.R. Ranganathan was the Chairman of the Committee. The responsibilities of the committee were to recommend standards for teaching, examination and research in library schools. The UGC accepted all the recommendations of the Review Committee on Library Science (1961) for implementation by the library schools. The Review Committee on Library Science report entitled "Library Science in Indian Universities" was published by UGC in 1965.

**The Education Commission (1964-66)**

In July 1964, the Government of India appointed the Education Commission under the Chairmanship of Prof. D. S. Kothari to evaluate the education system and the libraries. This Commission is known as the Kothari Commission (1964-66). The Report of the Education Commission (1964-66) was published under the title "Education and National Development". Kothari Commission also suggested monetary guidelines. "As a norm, a university should spend each year about Rs. 25.00 for each student and Rs. 300.00 per teacher or from 6.5% to 10% of the total budget depending on the stage of development of each university library."

The Education Commission (1964-66) was unique in several respects and had several special features. Of these, two deserve special notice here, viz. (1) its comprehensive approach to educational reconstruction; and (2) its attempt to project a blueprint of a national system of education for India (Naik,1982).

### Mehrotra Committee (1983)

The UGC appointed a Committee on December 24, 1983 under Professor R.C. Mehrotra's Chairmanship, to consider the revision of pay scales, career/ professional development and service conditions of teachers in the universities and colleges. Mehrotra Committee considered the revision of pay scales of Librarians and Directors of Physical Education too. The Committee submitted the report on the revision of pay scales of Librarians to the UGC on September 3, 1986.

The Government of India accepted the recommendations of the Mehrotra Committee and issued orders to implement revised pay scales and qualifications for teachers in universities and colleges. The recommendations regarding scales of pay, career advancement, promotion, terms and conditions of revised scales, recruitment and qualification requirements, continuing education and appraisal of performance, service conditions including period of probation, non-vacation, age of superannuation, code of professional ethics, grievances redressal mechanism and pay fixation formula were to be modified suitably for the librarians at different levels.

### Committee on National Network System for Universities (1988)

In 1988, the UGC constituted a committee on the National Network System under the Chairmanship of Prof. Yash Pal, the then UGC Chairman, to suggest measures for networking of libraries and information centres in universities, deemed universities, institutions of national importance, UGC Information Centres, R&D institutions and colleges. The main objective of the Committee was to share the existing resources to optimize utilization and avoid duplications of holdings to have access to a wide range of literature.

In the first meeting on April 24, 1988, the Committee constituted a Working Group to prepare a project report on 'Information and Library Network'(INFLIBNET) during the eighth five-year plan period. The members of the Working Group were experts and officials of some representative organizations, libraries, information centers, computer and communication fields, etc. They commenced their work on June 2, 1988, and completed the assignment on August 31, 1988.

**Curriculum Committee (1990-93)**

In September 1990, the Curriculum Development Committee (CDC) on Library and Information Science (LIS) was constituted on the recommendation of the UGC Panel on Library and Information Science to modernize the curriculum of LIS courses. The members of the Committee were Prof P.N. Kaula, Prof. S.G. Mahajan, Prof. (Ms.) A.K. Anand, Dr. P.K. Mahapatra, Dr I.K. Ravi Chandra Rao, Prof. & Head, Department of Library Science of Aligarh Muslim University or his representative, and Dr. O.P. Gupta of the UGC.

**INFLIBNET (1988)**

Information and Library Network (INFLIBNET) Centre, Gandhinagar is an Autonomous Inter-University Centre (IUC) of the University Grants Commission, New Delhi (Ministry of Education, Govt. of India). It is a major National Programme initiated by the UGC in March 1991 as a project under the IUCAA, and it became an independent Inter-University Centre in June 1996. INFLIBNET is involved in modernizing university libraries in India using state-of-art technologies for the optimum utilization of information. INFLIBNET is set out to be a major player in promoting scholarly communication among academicians and researchers in India.

**Major Programmes of INFLIBNET Centre are**

1. INDCAT: Online Union Catalogue of Indian Universities
2. SOUL: Online Union Catalogue of Indian Universities
3. e-ShodSindhu: Consortium for Higher Education e-Resources
4. Shodganga: A Reservoir of Indian Theses

**Major projects of INFLIBNET Centre are**

1. Shodh Shuddhi: Provides Plagiarism Detection Software to Institutions
2. IRINS: Research Information Management System for Institutions
3. Vidya-Mitra: Integrated e-Content Portal
4. Vidwan: Expert Database & National Researcher Network

**Rastogi Committee (1997- 1998)**

The UGC appointed a committee in 1997 under the Chairmanship of Prof. R. P. Rastogi to consider the revision of pay scales and career advancement of the academic staff of the universities and colleges. On the recommendations of the Rastogi Committee, the Ministry of Human Resource Development announced the revised pay scales of the teachers, librarians and physical instructors. The Ministry issued subsequent clarifications while retaining the parity of librarians' pay scales with teachers. It withdrew the parity in career advancement, re-employment or extension of service and fixation of pay scales, etc.

## Curriculum Development Committee (1997-2001)

The UGC constituted subject panels in different disciplines, including for Library and Information Science (LIS) courses, with Professor C. R. Karisidappa as its convener to formulate a model curriculum and prepare the status report with modifications of existing guidelines necessary. The Panel of LIS had its first meeting in Karnataka University, Dharwad, on March 1998, after meeting conveners of all disciplines with Dr. Hari Gautam, Chairman, UGC, on July 24, 1997. The plan for preparing the Status Report and Model Syllabus in LIS was discussed in a couple of panel meetings. The Committee surveyed LIS education in India to collect information regarding the current syllabus and other details by circulating a questionnaire to all the Department of LIS. Four separate Zonal Meetings were held at Kurukshetra University, Utkal University, Osmania University and North Eastern Hill University. Meanwhile, the Chairman advised constituting three groups of eight members each for the preparation of the Curriculum Development Committee Report. These three groups finalized courses, modules, model syllabus, mark pattern, teaching hours, and credits for every course.

## N-LIST of INFLIBNET

Academic libraries in India have a mixed collection of print and electronic resources. Access to online services is valued. Libraries are reaching their users through remote access tools. The resources include E-books, online databases, e-journals, online journal archives, Institutional Repositories (IRs), Electronic Thesis and Dissertations (ETD) etc. Universities are also purchasing or subscribing to e-resources through consortia or individual capacity. ILMS (Integrated Library Management Software) are available to automate libraries.

Other functions and roles of UGC in higher education

1. UGC and academic libraries.
2. UGC and library committee
3. UGC and scales of library staff

4. UGC and library science training
5. UGC and computers
6. UGC and reprographics service

**Higher Education Commission of India, (HECI)**

Higher Education Commission of India, (HECI) Bill 2019 was passed in the Parliament to create a single regulator for higher education, replacing the University Grants Commission (UGC) and the All-India Council for Technical Education (AICTE) (Economic Review,2020,2021).

**National Education Policy 2020**

The NEP 2020 is based on the Draft National Education Policy 2019, formulated under the chairmanship of Dr K Kasturirangan, former chairman of the Indian Space Research Organisation. The Policy was approved on July 29, 2020, by the Union Cabinet. The NEP (2020) is a paradigm shift in the structure of education.

To ensure vibrant libraries in communities and educational institutions, it will be imperative to make available adequate library staff and also devise appropriate career pathways and CPD for them (National Educational Policy 2020)

**Conclusion**

There has been a significant development in the country's higher education since the dawn of independence. The chapter contains the genesis and development of libraries in India with special reference to the university libraries in India since Independence. Different committees and commissions strongly recommended the importance of libraries in university education. In the present scenario, a single library cannot meet the users' demand for all information. Hence, libraries prioritize pooling, sharing, and optimizing resources using modern technology and a consortia-based collective approach.

**References**

Economic Review 2020. (2021). Vol. I. Thiruvananthapuram: Kerala State Planning Board. Retrieved from https://spb.kerala.gov.in/sites/default/files/2021-01/English-Vol-1_0.pdf

Ghosh, Suresh C (2007). History of Education in India. Jaipur: Rawat Publication, p. 452.

https://inflibnet.ac.in

https:// old.mu.ac.in.

https://www.culibrary.ac.in

https://www.unom.ac.in/

India: (2020). National Education Policy 2020. New Delhi: Ministry of Human Resource Development.p.52

Kaula, P.N. (Ed.) (1958). Library Movement in India. Delhi: Delhi Library Association, 1958.p.58.

Naik, J P (1982). The Education Commission and After. New Delhi: Allied Publishers Private Limited.p.1

Swayam, "Emerging Trends & Technologies in Library & Information Services (ETTLIS)," https://swayam.gov.in/nd2_arp19_ap78/preview.

# 4

# Quality of Work Life An Overview

## Introduction

Quality of Work Life (QWL) is a popular concept among the organizations. Work is an essential part of everyday human life. Work influences the overall quality of life. Human resources are one of the main assets of any organization. Quality of work life (QWL) is one of the important subsystems of human resource management. There are economic, social, psychological and organizational aspects of work life. Many new subsystems are being added to human resource management with employees' changing needs, culture, expectations, and work environment. QWL is one such new area. Quality of Work Life describes experience related to the employment of a person. QWL has much to do with developing the organization and progress of human resources in both public sector and private sector organizations. Quality of Work Life refers to favourableness or unfavourableness of a job for personnel working in the organization. Human resource management is the core of effective service in university libraries. QWL affects the performance level of employees and managerial skills in organizations.

Principles of scientific management developed by F.W. Taylor. It created a new awareness regarding human resources. The idea of QWL was conceived in the 1970s. The term "Quality of Work Life" was introduced in the first international labour conference on QWL in Arden House, Columbia, New York, in 1972. The International Council for Quality of Working Life (ICQWL) was formed to facilitate research on and action for quality of working life. The conference dealt with the practice and theory of democratization of workplace. The conference acknowledged the necessity to build up a solid theoretical corpus in the area of QWL research. Thus, the International Council for the Quality of Working Life was created in 1973 to promote research and exchange information concerning mental health at work. The first open international conference was organized in Toronto in August 1981 on Quality of work life.

Richard E. Walton (1975) of the USA advocated eight broad criteria for measuring a desirable QWL. The criteria are adequate and fair compensation, safe and healthy environment, opportunity to use and develop human capacities, opportunity for career growth, constitutionalism, work, quality of life and social relevance.

Lawler (1975) proposed the following four characteristics to include in measuring QWL. The four characteristics are:

1) It must measure the important aspects of QWL.
2) It must also have sufficient face validity in the eyes of anyone likely to use it.
3) It must be objective and, consequently, verifiable without any possibility of being manipulated.
4) It must be capable of distinguishing between individual differences within the same work environment.

The term "quality of work-life" means different things to different persons. For example, to a worker on an assembly line, it may just mean a fair day's pay, safe working conditions, and a supervisor who treats him with dignity. To a new entrant, it may mean opportunities for advancement, creative tasks, and a successful career (Raja et al.,2010)

Quality of work life includes several motivational factors. Frederick Herzberg emphasized it in his motivation theory. Job enlargement and job rotation are the part of these motivators. Job enrichment helps the growth of employee. Job rotation, job enlargement, job enrichment improves the efficiency of employees in organizations.

## Quality of Work Life and Theories of Motivation

QWL is a multi-faceted concept to improve organizational efficiency. There exists a relationship between QWL and productivity in the work environment. Quality of work Life is a potential for library and information domain. Theories relating to motivation has created a sound base for Quality of Work Life.

## Maslow's Hierarchy of Needs

Abraham Maslow (1943) is an American Psychologist known for creating Maslow's Hierarchy of Needs. The idea of Maslow was fully expressed in 1954. The study predicated on fulfilling innate human needs in priority. Maslow described human needs as being relatively fluid, with many needs being present in a person simultaneously. Human needs identified by Maslow are in five hierarchies. Maslow's hierarchy of needs is portrayed in the shape of a pyramid. At the bottom of the hierarchy are the Basic needs or Physiological needs. The next level is Safety Needs. These two levels are together called basic needs, and it is important to the physical survival. The third level of need is Love and Belonging. The fourth level is the Esteem level. The third and fourth level of need is together called psychological needs. The fifth is the self-actualization level. It is also called self-fulfilment needs. The need for Self-actualization occurs when individuals reach a state of harmony and understanding because they are engaged in achieving their full potential. The self-actualization state focuses on builds their image.

### Herzberg's Motivation-Hygiene Theory (Two Factor Theory)

The two-factor theory was developed by an American Psychologist, Frederick Herzberg (1959). It is a new dimension to Maslow's need theory of hierarchy. Herzberg is famous for introducing "Job enrichment and Motivation Hygiene Theory". The two-factor theory is also known as dual-factor theory and Herzberg's motivation-hygiene theory. The two-factor theory states that "there are certain factors in the workplace that cause job satisfaction while a separate set of factors cause dissatisfaction, all of which act independently of each other". The two-factor theory distinguish between motivators and hygienic factors. Motivators give positive satisfaction because it is arising from the intrinsic condition of the job. Hygienic factors do not give positive satisfaction or higher motivation because they arouse from extrinsic to job.

### McGregor's Theory X and Theory Y

Douglas McGregor proposed Theory X and Theory Y in 1950. The two contrasting theories are the theories on human work motivation and management. Theory X explains "the importance of heightened supervision, external rewards, and penalties". Theory Y highlights "the motivating role of job satisfaction and encourages workers to approach tasks without direct supervision". "For McGregor, Theory X and Theory Y are not opposite ends of the same continuum, but rather two different continua in themselves. To achieve the most efficient production, a combination of both theories may be appropriate".

### Job Rotation

Job rotation is a management approach of authorities to rotate the employees in their jobs. It promotes flexibility, experience and variety in the job of employees in an organization.

**Advantages of Job Rotation**

- Alleviation of stresses of work at the same position
- Reduction in risk factors
- Reduction in boredom and monotony
- Development of multiple skill
- Job security

**Disadvantages of Job Rotation**

- Cost involved to train the workers
- Specialization
- Resistance from employees

**Job Enrichment**

Herzberg (1968) developed the concept of job enrichment. Job enrichment, a motivational tool of personnel, is a vertical loading of the job. It increases level of duty and authority in a job. It makes the job more creative and also increasing the duties.

**Advantages of Job Enrichment**

- Learn new skills
- Reduce boredom:
- Creates a positive work environment
- Maximise productivity

**Disadvantages of Job Enrichment**

- Lower productivity
- Heavier workload
- Clash with non-participants
- Poor performance

**Job Enlargement**

Job enlargement increases the scope of the job and motivates employees in the organization. Job enlargement extends the range of jobs within the same level and adding to the current level. Hulin and Blood (1968) defined job enlargement as the process of allowing individual workers to determine their own pace within limits. Herzberg (1968), an American psychologist, introduced job enlargement and the Motivator-Hygiene theory. He referred to job enlargement as a horizontal loading of a job.

**Advantages of Job Enlargement**

- Increase skills in a new field
- Improve earning capacity
- Ability to learn a wide range of activities
- Reduction of cost

**Disadvantages of Job Enlargement**

- Demotivation of employees
- Dissatisfaction among the personnel
- Increase in workload

Various motivation factors increasing the efficiency of employees and also encourage job satisfaction of employees in organizations. No doubt, job rotation, job enrichment, and job enlargement is helping the growth of employees and contribute more towards achieving the organisation's goal. In this study, an attempt has been made to assess the quality of work-life, job enrichment and job enlargement prevailing in the university libraries of Kerala.

## Quality of Work Life in Libraries

Quality of Work Life has been discussed in different disciplines, including Library and Information Science. Library and Information service is essential for the functioning of the universities. Libraries support the teaching, learning, research and extension activities of the university. In service institutions like libraries and Information centres, the role of human resources is important. Libraries and information centres are continually adopting changes in connection with developments in the field of management. LIS professionals of the universities are the inherent part of the teaching, research and extension activities. For a library professional, quality of work life refers to the totality of their feelings and experiences of life at work in a library. The quality of work life of employees working in different universities can be enhanced by adopting the constructs of quality of work life. Some of the constructs of quality of work life are adequate, fair compensation, safe and healthy working conditions, immediate opportunities to use and develop human capacity, opportunities for continued growth and security, social integration in the work organization, constitutionalism in the work organization, the social relevance of working life, work and total life space, etc. It has generally been agreed however that quality of working Life is conceptually similar to well-being of employees but differs from job satisfaction which solely represents the workplace domain (Lawler, 2005).

Walton defines the quality of work life as a process by which an organization responds to employees' needs for developing mechanisms to allow them to share fully in making the decisions that design their lives at work (Walton, 1973).

Satisfaction of clientele in a library is linked with job satisfaction of staff. Satisfaction of clientele cannot be achieved unless the staff of library enjoys good quality of work life. We have qualified library professionals to manage our libraries. But there is a lack of initiative and drive on the part of authorities to pay enough attention to the employees' personal needs and motivational factors.

### Job Enrichment in Libraries

Frederick Herzberg developed job enrichment as a concept in 1968. Job enrichment is one of the motivational techniques in modern management. It involves combining various existing and new tasks into one large module of work. The advantages of job enrichment are autonomy, the chance to learn new skills and gain knowledge through work experience other than a regular job, more variety and responsibilities in the job. It also encourages a positive environment, recognition and job motivation. The negative outcome of job enrichment is that the employee may not like to accept new tasks.

Job enrichment assigns responsibilities generally reserved for a higher level of employees. It allows complete autonomy to plan and carry out activities on the job. Job enrichment is a managerial motivational technique in the world of management. It cares about an employee as a person. It is a vertical job restructuring technique where the focus is on giving the employee more authority, independence, and control. Job enrichment creates a desire for an employee to contribute in the form of hard work, loyalty and dedication. It helps employees to learn new skills, reduce boredom and creates a better work environment.

### Job Enlargement in Libraries

The role of library and information science professionals has changed in the modern technological era. Job enlargement is one of the major techniques of job design. It is a management technique in human resource management, policy-making and job motivation in the modern era. The job of the librarian has changed from curator to custodian to facilitator. The library and information science professionals are expected to perform additional tasks in addition to their regular duties. It is known as job enlargement.

Job enlargement is defined as the process of allowing individual workers to determine their own pace (within limits), to serve as their inspectors by giving them responsibility for quality control, to repair their own mistakes, to be responsible for their machine set-up and repair, and to attain choice of method. (Hulin and Blood,1968) It is referred it to the addition of interrelated tasks as 'horizontal job loading'(Herzberg, 1968). Job enlargement is a horizontal restructuring technique where the focus is increasing the number of assignments but does not change the overall authority, autonomy, and control of the projects. It is just an increase in the work of an employee without assigning any responsibility.

Job enlargement extends the scope of duties and responsibilities in quantitative nature. It leads to job satisfaction, higher performance in the job. It creates an opportunity to perform various jobs at the same level in the library and information centres. Job enlargement has a positive outcome and motivational impact, and it may have a negative impact that is likely to be resisted by employees.

## Conclusion

QWL can accomplish the well-defined objectives of the organization effectively. QWL varies from organization to organization. QWL motivate employees to participate in the decision-making process and enhance the efficiency of the organization. LIS professionals are instrumental in helping LIS professionals with sincerity and timely accomplishments. Their efforts are praiseworthy therefore appreciated. It is time to equip LIS professionals to survive in a changing scenario. The professionals will adopt changes and suitably adapt themselves to improve their performance and to rise to the occasion. The concerted efforts of management and employees strengthen organization by transcending all barriers.

## References

Davis, L. & Cherns, A. (Eds.). (1975). The quality of working life. New York: Free Press.

Hattangadi, Vidya. (2015). Theory X & Theory Y. International Journal of Recent Research Aspects. 2: 20–21.

Herzberg, F. (1968). One more time: How do you motivate employees. Boston: Harvard Business Review. pp. 46–57.

Herzberg, Frederick., Mausner, Bernard & Snyderman, Barbara B. (1959). The Motivation to Work (2nd ed.). New York: John Wiley.

Hulin, Charles L., & Blood, Milton. R. (1968). Job enlargement, individual differences, and worker responses. Psychological Bulletin, 69(1), 41–55.

Lawler, E. E. (1975). Measuring the psychological quality of working life: The why and how of it. The quality of working life, 1, 123-133.

Lawler, E.E. (2005). Strategies for improving the quality of work life, American psychologist, 37, 486-493.

Maslow, A (1954). Motivation and personality. New York, NY: Harper.

Maslow, A. H. (1943). A theory of human motivation. Psychological Review, 50 (4): 370–396.

Raja E A., Prabhu, N R V., & RAO, P K (2010). Managerial behaviour and effectiveness (P.116). New Delhi: Excel Books.

Walton, R. E. (1975). Criteria for Quality of Working Life. In: Davis, L.E., Cherns, A.B. and Associates (Eds.) The Quality of Working Life, The Free Press; 1975. P.91-104.

Walton, Richard. (1973), Quality of work life: What is it? Sloan Management Review. 15(1), 11-21.

# 5

# Profile of Selected University Libraries in Kerala

## Introduction

Libraries are a center of information for the academic community of the universities in Kerala. It acts as a gateway to global knowledge. Libraries provide seamless access to the information that supports teaching, learning research, and extension. It also ensures fair access to knowledge in the academic community. The libraries fulfill the objectives of their parent organization. It caters to the educational needs of students, researchers, and faculty/Scientists of the universities for study, research, teaching, and extension activities. Now, users can access resources, services, and updates of the library 24/7 on their cell phones. The users can visit the online library whenever they want. Libraries are adopting technology to provide all possible services to students, research scholars, faculty members, and scientists.

Kerala can quickly notice the latest global trends in Library and Information Science (LIS) and quickly adopt them. The organizational structure of university libraries in Kerala consists of University Library or Central Library, Departmental libraries, Study center libraries, and constituent college libraries. The volume, collection, and services of the libraries differ from library to library, as well as the utilization of its resources. In this chapter, an attempt has been made to provide a short profile of Kerala's selected eight state university libraries. State universities in Kerala are established under Acts passed by the state legislature. There are 14 state universities functioning in Kerala (Economic Review 2020). The chapter contains the origin and growth of universities and their libraries and qualifications for University Assistant Librarian / College Librarian, University Deputy Librarian, and University Librarian as per UGC regulation, 2018. It also deals with recruiting LIS professionals in the universities of Kerala.

## Kerala University Library

### The University

It was established as the University of Travancore by the Travancore University Act in 1937, reconstituted as the University of Kerala by the Kerala University Act of 1957, and presently governed by the Kerala University Act of 1974. The University of Kerala is known as the mother university of the state. It is one of the first sixteen universities in India. The university ranked 22nd among universities in India in NIRF 2019 and 49th among universities in India in the QS World Ranking. The jurisdiction of the University of Kerala is to Thiruvananthapuram, Kollam, Alappuzha districts, and some parts of Pathanamthitta district.

The University of Kerala has two major campuses, the largest being the Kariavattom Campus. The Palayam/Senate Hall Campus is in the middle of the city. The University of Kerala has its teaching and research departments. The forty-three departments of teaching and research are organized as eleven schools. Teaching, research, and Knowledge extension are the mandates of the departments. They primarily focus on post-graduate (masters) programmes, M. Phil. Programmes, doctoral (Ph.D.) programmes and post-doctoral programmes. Affiliated colleges run a large number of bachelor's programmes and a lesser number of master's and doctoral programmes. The University has also established 10 University Colleges of Teacher Education (KUCTE) and 35 University Institutes of Technology (UIT), both of which offer undergraduate programmes (B.Ed. in KUCTE's and BSc (Computer Science/IT/ Electronics), B. Com. and BBA in UITs). However, masters programmes are available in select UITs. The University College of Engineering at Karyavattom offers Engineering education at the undergraduate level. The University has over 189 affiliated colleges. Thirty-two colleges offer postgraduate teaching programs. Some affiliated colleges offer M.Phil. programs, and some have been recognized as research centers. In addition to teaching departments, the University also has several centers for studies and research in specialized areas. The University has also instituted three chairs.

Department of Library and Information Science was established at the University of Kerala in 1961. The university had a Library Science faculty member who is a Prof. S.R. Ranganathan student. The department offers MLISc (Master of Library and Information Science), Ph. D., and Post Doc in Library and Information Science.

### Library

Kerala University Library was established in 1942 to provide teaching and learning support to the academic community. Kerala University Library System has over eight lakhs of books. The University has a Central Library (Kerala University Library-KUL) near its city campus and caters to the knowledge quest of all students, staff, and the public. In addition to KUL (over 3,30,000 books), the Karyavattom campus has a Campus Library. Department libraries (above 5,00,000 books) and College libraries (each with above 20,000 – 1,00,000 books) provide information services to students. It has rare books and bibliographies in its collection. The annual growth rate of stock in the library is 1.53%. The different sections of the Kerala University Library are:

1. Acquisition Section
2. Circulation Section
3. Documentation and Information Services Section
4. Information Technology Section and E-resource Centre.
5. Kerala Reference Section
6. Maintenance Section
7. Periodical section
8. Reference Section
9. Research Section
10. Technical Section
11. United Nations and World Bank (Depositories) Section

## Existing Cadre and Pay structure of LIS professionals at the University of Kerala

Table 5.1 presents the cadre and scale of pay of LIS professionals working in the University of Kerala.

**Table 5.1:** Cadre and Scale of Pay of LIS Professionals at University of Kerala

| Sl.No. | Cadre | Scale of Pay |
|---|---|---|
| 1 | University Librarian | UGC |
| 2 | Deputy Librarian | |
| 3 | Assistant Librarian | |
| 4 | Deputy Librarian (Non-UGC) | 77200-140500 |
| 5 | Assistant Librarian Grade I (Non-UGC) | 59300-129000 |
| 6 | Assistant Librarian Grade-II (Non-UGC) | 55200—115300 |
| 7 | Reference Assistant | 50200-105300 |
| 8 | Technical Assistant (Library) | 45600-95600 |
| 9 | Library Assistant | 39300-83000 |

## University of Calicut

### The University

The University of Calicut, the largest University in Kerala, was established in 1968. The University aims to nurture excellence in education and research in its catchment areas of Northern Kerala. The university emphasizes fostering quality human resources and promoting productive research that benefits local communities and wider humanity. The University was created through a government plan bifurcating Kerala University, by which, the four post-graduate departments of the University of Kerala operating in Calicut were annexed to the new University along with 54 constituent colleges spread across seven northern districts. The university has 34 postgraduate and research departments, three centers, 11 chairs, and 395 affiliated colleges.

Department of Library and Information Science at the University of Calicut was established in 1985. The department now offers a two-year Master of Library and Information Science (M.Lib.I.Sc.) course. The department has also launched M. Phil. and Ph.D. programs in Library and Information Science.

### University of Calicut Library (CHMK Library)

C.H. Mohammed Koya Library is the Central Library of Calicut University. The library's total collection includes books, gift books, theses/dissertations, and bound volumes. The collection also contains CD ROMs and Microfilms on various topics. University faculty members, research scholars, and students can access e-resources from any location provided through the Shibboleth Login facilitated by INFED.

The various sections of the library are:

Acquisition Division

Technical Division

Maintenance Division

Circulation Division

Periodical Division

Reference Division

Photocopying and Binding Section

Digital Library

ICT Division/Research Desk

Services available in the library are Plagiarism checks, citation management, Book Lending Service, Reference Services, Information Services, Compilation of Bibliographies, New Additions List, and Reprographic Service.

### Existing Cadre and Pay structure of LIS professionals at the University of Calicut

Table 5.2 presents the cadre and scale of pay of LIS professionals working at the University of Calicut.

**Table 5.2:** Cadre and scale of pay of LIS professionals at the University of Calicut

| Sl.No. | Cadre | Scale of pay |
|---|---|---|
| 1 | Information Scientists | UGC |
| 2 | University Librarian | |
| 3 | Deputy Librarian | |
| 4 | Junior Librarian | 50200-105300 |
| 5 | Professional Assistant Grade I | 45600-95600 |
| 6 | Professional Assistant Grade II | 39300-83000 |

## Mahatma Gandhi University, Kerala

### The University

The Mahatma Gandhi University was established on 2nd October 1983, and its jurisdiction extends to the Revenue Districts of Kottayam, Ernakulam, Idukki, and parts of Pathanamthitta and Alappuzha. The University conducts a range of programmes at the Undergraduate, Postgraduate, M.Phil. and Doctoral levels through its 17 University Departments, 1 International and Inter-University Centre, 7 Inter-University Centres, 10 Inter-School Centres, 77 Government/ Aided Affiliated Colleges including 10 Autonomous Colleges (of which 8 are Colleges with Potential for Excellence), 200 Unaided Affiliated Colleges and 273 Recognized Research Centres. It imparts education in the interdisciplinary and conventional disciplines of Science, Social Science, and Humanities and the professional disciplines of Engineering, Technology, Legal Studies, Pedagogy, Pharmacy, and Nursing.

### Mahatma Gandhi University, Kerala Library

The University Library was started in 1989. University Library provides a variety of learning environments. These include individual workspaces, rooms for group study and teaching, audiovisual access, and an online information retrieval system. The library has taken exclusive initiatives to imbibe ICT into the entire gamut of its activities. MGU Library and Information System consists of University Central Library, more than 30 libraries of the departments/schools, and 4 Study Centres on different campuses.

The library is a member of the INFLIBNET Centre, Ahmedabad (Information and Library Network), an Inter-University Centre of the University Grants Commission. INFLIBNET Centre support has enabled it to improve the quality and relevance of university information services. The University Library is a fully automated system. It is also a member of the DELNET (Developing Library Network). As a member of this network, the library provides its users access to the resources of other major libraries in the country.

Mahatma Gandhi University comes under the UGC-INFONET Programme. Through the UGC-INFONET E-journals Consortium, the Library has access to e-journals and databases, including SciFinder Scholar and Web of Science. The library also has access to major online databases, such as EBSCO, ProQuest Dissertations and Theses, Oxford Scholarship Online, IEEE All Society Periodicals Package, etc. With the Mahatma Gandhi University Online Theses Digital Library launching, Mahatma Gandhi University has become the first Indian University to adopt Open Access of its doctoral research theses to academics worldwide through www.mgutheses.in

The University Library has a collection of books, e-books, theses, and bound volumes of journals. In addition to this, the library subscribes to current Indian and Foreign journals, Online Databases, and Online Journal Archives. The total collection of books in the university library, Department/Center Libraries, and Study Center Libraries is around two lakhs.

## Existing Cadre and Pay Structure of LIS professionals in the Mahatma Gandhi University, Kerala Library

Table 5.3 presents the cadre and scale of pay of LIS professionals working at Mahatma Gandhi University, Kerala.

**Table 5.3:** Cadre and scale of pay of LIS professionals in the Mahatma Gandhi University, Kerala

| Sl.No. | Cadre | Scale of pay |
|---|---|---|
| 1 | University Librarian | UGC |
| 2 | Deputy Librarian | |
| 3 | Assistant Librarian | |
| 4 | Assistant Librarian Grade I (Non-UGC) | 59300-120900 |
| 5 | Assistant Librarian Grade-II (Non-UGC) | 55200-115300 |
| 6 | Reference Assistant | 50200-105300 |
| 7 | Technical Assistant Library | 45600-95600 |
| 8 | Library Assistant | 39300-83000 |

## Cochin University of Science and Technology (CUSAT), Kochi

### The University

The University of Cochin, established in 1971, stands as the symbol of commitment from the Government of Kerala towards the new scientific policy resolution adopted by the Central Government on 4th March 1958. The University of Cochin was reorganized and converted into a full-fledged University of Science and Technology by the Cochin University of Science and Technology Act, 1986 to promote graduate and post-graduate studies and advanced research in Applied Sciences, Technology, Industry, Commerce, Management, and Social Science.

### Library

The Cochin University Library was established in 1975. It is a central agency for meeting the information requirements of the university's academic community and for the collection, organization, and dissemination of information. The library has been exploring all the possibilities and advancements of technological applications to transform itself into a modern library. Located centrally in the campus, the library resources are open to all the university departments. Currently, the library collection exceeds 95,000, including several valuable reference books, doctoral theses, and back volumes of periodicals, patents, CD-ROMs, etc.

The library offers Loan service, Inter-Library Loan, Literature Search, Compilation of Bibliographies, Reprographic Service, Conference Alert Service, New Addition Alert Service, and Consultancy Service. The library has campus-wide- IP-enabled access to various electronic databases, e-journals, and online books.

The library provides access to print periodicals, online journals, databases, and e-books. The Academic community of CUSAT can access e-books, thesis databases, journal databases, e-journals, and other e-resources. Online access is also provided to citation and scientometric databases.

The library also facilitates access to several electronic journals by participating in the consortia -E- ShodhSindhu. It is a project initiated by the Ministry of Human Resource Development (MHRD) and executed by INFLIBNET. Under this consortia, electronic full-text access is provided to the scholarly literature of all major areas of learning. Members can avail of DELNET, UGC-INFONET, and the JCCC portal (an e-journal gateway to Indian Universities and colleges), which also provides an interlibrary loan between various Universities in India.

The library has shifted to an open-source koha software for library automation. The bibliographic records of books, journals and theses in the library can now be accessed globally through the web OPAC facility. The library has also developed the Union catalog of all the libraries in CUSAT. The library houses an Institutional Repository. A research gateway to publications of CUSAT faculty is provided through YUJ, a bibliographic information system. A media repository is a library of various events, lectures, etc., conducted in the University.

### Existing Cadre and pay Structure of LIS professionals in CUSAT

Table 5.4 presents the cadre and scale of pay of LIS professionals working at the Cochin University of Science and Technology

**Table 5.4:** Cadre and scale of pay of LIS professionals in CUSAT

| Sl.No. | cadre | Scale of pay |
|---|---|---|
| 1 | University Librarian | UGC |
| 2 | Deputy Librarian | |
| 3 | Information Scientist | |
| 4 | Assistant Librarian | |
| 5 | Assistant Librarian | 55200-115300 |
| 6 | Junior Librarian | 50200-105300 |
| 7 | Professional Assistant Grade I | 45600-95600 |
| 8 | Professional Assistant Grade II | 39300-83000 |

## Sree Sankaracharya University of Sanskrit, Kalady, Ernakulam

### The University

The Sree Sankaracharya University of Sanskrit (SSUS) is one of the pioneering Sanskrit Universities in India in teaching, research, and innovations. The University was established in 1994 in Kalady, Kerala. The foundational aim of the University is to impart knowledge in Sanskrit, other Indian and Foreign Languages, Social Sciences, and Fine Arts. As an institution imbibing multiple cultural heritages and housing traditional and modern knowledge systems and art forms, the University envisions a harmonious societal advance through educational excellence. The Sree Sankaracharya University of Sanskrit has eight regional campuses across Kerala.

### Library

University Library offers book lending, reference, IT-related information, user education, Web OPAC, photocopying, newspaper, and bindery services. The library's print resources include Indian journals and foreign journals. The library also has access to Shodh Ganga, e-Shodh Sindhu-UGC INFONET, Indcat, and E-Vidwan. The library's free and open-access resources include E-Journals,

E-Books, Electronic Thesis & Dissertation, Audio Books E-Reference, Dictionaries, Digital Libraries, and e-Learning. Periodicals of the library contain E-newspaper and magazines. The library published an International Research Journal SCAIRIRJ (Saraswatham Centre for Advanced Information Resources International Research Journal). It's a biannual journal with multidisciplinary coverage. There is a Remote Access facility to E-Resources Through INFLIBNET INFED.

The library has the following sections:

1. Circulation Section
2. Acquisition Section
3. Technical Section
4. Reference Section
5. INFONET & Digital Library
6. Periodical Section
7. Journal Section
8. Manuscript Section
9. Newspaper Reading Room
10. Research Section
11. Maintenance Section (Stack Room)
12. Reprographic Section

### Existing Cadre and pay structure of LIS professionals in SSUS

Table 5.5 presents the cadre and scale of pay of LIS professionals working in the Sree Sankaracharya University of Sanskrit.

**Table 5.5:** Cadre and scale of pay of LIS professionals in SSUS

| Sl.No | cadre | Scale of pay |
|---|---|---|
| 1 | Deputy Librarian | UGC |
| 2 | Reference Assistant | 50200-105300 |
| 3 | Technical Assistant | 45600-95600 |
| 4 | Library Assistant | 39300-83000 |

## Kerala Agricultural University, Thrissur

### The University

The Kerala Agricultural University was established in 1971. The university imparts education and undertakes research and extension activities in agriculture and allied sciences. The University fulfills its mission through a network of institutions spread over the State. It comprises eight constituent colleges, six Regional Agricultural Research Stations (RARS), sixteen research stations, one Central Training Institute, one Communication Centre, and seven Krishi Vigyan Kendras (KVK). There is a network of libraries, including a Central Library on the main campus, constituent colleges, Regional Agricultural Research Stations (RARS), research stations, and Krishi Vigyan Kendras (KVKs). The University offers diploma, graduate, post-graduate and doctoral programmes in almost all Agriculture, Agricultural Engineering, Forestry, Horticulture and allied sciences disciplines.

### Library

The Kerala Agricultural University has a network of libraries, including a Central Library, constituent colleges, RARS, research stations, and KVKs. The Kerala Agricultural University Library and Information System (KAULIS) was established in 1995. The KAULIS consists of the Central Library, college libraries in constituent colleges, and research stations. It has digitized a repository of information on agriculture, including Crop and Animal Production, Fisheries, Forestry, Nutrition, and Rural Development. It facilitates digital access to the most advanced sources of information like CABI e-Books (https://www.cabi.org), CAB Abstracts, KrishiPrabha, Indiastat.com, etc., and facilitates knowledge updation for students and faculty.

Consortium for e-Resources in Agriculture (CeRA) provides access to the full text of journals from reputed publishers like Elsevier, Springer, Annual Review Inc., Wiley, etc. Apart from that, users can request articles from any of the libraries in the consortium through the Document Delivery Service. India Agristat, e-books, and e-book series published by Elsevier were also provided through CeRA. Krishikosh is a digital repository of accumulated knowledge in agriculture and allied sciences, which ICAR maintains. The full text of the thesis submitted to the University can be accessed through the Krishikosh. The University is an Institutional Member of DELNET (Developing Library Network), which provides access to bibliographic records of books, journals, articles, etc. The Inter-Library Loan (ILL) and Document Delivery Services (DDS) are two of the most popular services of DELNET. Indian Digital Ensemble of Agricultural Libraries (IDEAL) is a platform for sharing library holdings through the Union Catalogue (AgriCat) of the Agricultural University

Libraries of the Indian National Agricultural Research and Education System (NARES). Bibliographic details of more than one lakh books available in College Libraries and Central Libraries are available through IDEAL.

### Existing Cadre and Pay structure of LIS professionals in Kerala Agricultural University

Table 5.6 presents the cadre and scale of pay of LIS professionals working at the Kerala Agricultural University.

**Table 5.6:** Cadre and scale of pay of LIS professionals in KAU

| Sl.No. | cadre | Scale of pay |
|---|---|---|
| 1. | University Librarian | UGC |
| 2. | Deputy Librarian | |
| 3. | Assistant Librarian | |
| 4. | Reference Assistant Higher Grade / Assistant Librarian | 55200-115300 |
| 5. | Reference Assistant | 50200-105300 |
| 6. | Technical Assistant | 45600-95600 |
| 7. | Library Assistant | 39300-83000 |

## Kannur University, Kannur

### The University

Kannur University was established in 1996. The University, the name "Malabar University," had come into existence earlier by the promulgation of an Ordinance by the Governor of Kerala on 9th November 1995. The Kannur University was started with the objective of establishing a teaching, residential, and affiliating University, promoting the development of higher education in Kasargod and Kannur revenue Districts and the Mananthavady Taluk of Wayanad District in the State of Kerala. Kannur University is a multi-campus university with campuses at Kannur, Kasaragod, Mananthavady, Payyannur, Thalassery, Kanhangad, and other places necessary for providing study and research facilities to promote advanced knowledge in Science and Technology and other relevant disciplines.

The Department of Library and Information Science, Kannur University, was established in 2009. The Academic programmes of the department are M.Lib.I.Sc and Ph.D.

### Library

Kannur University Central Library was established in 1998 and opened to the academic community in 1999. The library was named Dr Hermann Gundert Central Library. The library serves the various sections of the university community. It is automated using the open-source software koha. In addition to Central Library, Campus libraries function on the different campuses at Kasargod, Nileswaram, Payyannur, Mangatuparamba, Thalassery, and Mananthavady. The library offers Lending services, Reference services, Interlibrary loans, access to DELNET, e-shodhsindu and e-shodhganga, Reprographic services, Conference alert service, Question bank services, Consultancy services, New addition alert services. The library collection contains books, periodicals, and newspapers. The university is subscribing to several databases. The library has CD-ROMs, theses and dissertations, and a special Thaliyola, Malabar, Career, and German languages collection. The library has e-journals and e-books.

### Existing Cadre and Pay structure of LIS professionals in Kannur University

**Table 5.7 presents the cadre and scale of pay of LIS professionals working at Kannur University.**

**Table 5.7:** Cadre and scale of pay of LIS professionals at Kannur University

| Sl.No | Cadre | Scale of Pay |
|---|---|---|
| 1 | University Librarian | UGC |
| 2 | Deputy Librarian | |
| 3 | Assistant Librarian | |
| 4 | Junior Librarian | 50200-105300 |
| 5 | Professional Assistant Grade I | 45600-95600 |
| 6 | Professional Assistant Grade II | 39300-83000 |

## Kerala Veterinary and Animal Science University, Pookode, Wayanad

### The University

The Government of Kerala established the Kerala Veterinary and Animal Sciences University (KVASU) in December 2010 to further education, research, and extension services in the field of Veterinary and Animal Sciences. The University encompasses three faculties- the Faculty of Veterinary and Animal Sciences, the Faculty of Dairy Science, and the Faculty of Poultry Science, and eight constituent colleges. The University has a vast network of instructional and research farms spread throughout the state.

### Library

Kerala Veterinary and Animal Science Library and Information System (KVASULIS) include libraries of the College of Veterinary and Animal Sciences at Mannuthy in Thrissur district and Pookode in Wayanad district, the College of Dairy Science and Technology located at Mannuthy, Pookode,

Thiruvananthapuram and Kolahalamedu, College of Avian Sciences and Management located at Thiruvazhamkunnu and College of Food Technology at Thumburmuzhy, Thrissur District, KVASULIS (University Library), VC's Library and Wildlife studies.

The library's primary mission is to support the educational and research programs of the Kerala Veterinary and Animal Sciences University by providing physical and intellectual access to information consistent with the institute's present and anticipated educational and research functions. The secondary mission is to serve as a resource center for the scholars and scientific community of the Veterinary and Animal Husbandry arena. This library offers services such as Web-oriented services, Online and Offline content searches, Reference and Referral Services, Circulation Services including Inter-Library Loan and Document Delivery Services, CAS and SDI, display of recent additions of periodicals and books, Literature search, Internet Browsing, Consultancy Service, User guidance and education etc.

Books are classified and arranged as per Dewey Decimal Classification (DDC). OPAC - Computerized Open Access Catalogue is available to locate the Documents. The web-enabled University Library has a vast collection of books and journals and online access to e-journals, viz. www. cera.jccc. in e-books: www.wileyonlinelibrary.com, open-source software (Institutional Repository), and Koha (Library management) software.

### Existing Cadre and pay structure of LIS professionals in Kerala Veterinary and Animal Sciences University

Table 5.8 presents the cadre and scale of pay of LIS professionals working at the Kerala Veterinary and Animal Sciences University.

**Table 5.8:** Cadre and scale of pay of LIS professionals in KVASU

| Sl.No. | Cadre | Scale of Pay |
|---|---|---|
| 1. | Assistant Librarian | UGC |
| 2. | Reference Officer HG | 55200-115300 |
| 3. | Reference Officer | 50200-105300 |
| 4. | Reference Assistant HG | 45600-95600 |
| 5. | Reference Assistant | 39300-83000 |

### UGC Regulation (2018)

UGC regulations on minimum qualifications for appointment of teachers and other academic staff in universities and colleges and measures for the maintenance of standards in higher education, 2018 have given the following minimum qualifications for direct recruitment to the posts of University

Assistant Librarian / College Librarian, University Deputy Librarian, and University Librarian.

**I. University Assistant Librarian / College Librarian**

i) A Master's Degree in Library Science, Information Science or Documentation Science or an equivalent professional degree, with at least 55% marks (or an equivalent grade in a point –scale, wherever the grading system is followed)

ii) A consistently good academic record, with knowledge of computerization of a library.

iii) Besides fulfilling the above qualifications, the candidate must have cleared the National Eligibility Test (NET) conducted by the UGC, CSIR or similar tests accredited by the UGC like SLET/SET or who are or have been awarded a PhD. Degree in accordance with the University Grants Commission (Minimum Standards and Procedure for Award of M.Phil./Ph.D. Degree) Regulations, 2009 or 2016 and their amendments from time to time as the case may be: Provided that the candidates registered for the Ph.D. degree prior to July 11, 2009, shall be governed by the provisions of the then existing Ordinances / Bye-laws / Regulations of the Institution awarding the degree, and such Ph.D. candidates shall be exempted from the requirement of NET/SLET/SET for recruitment and appointment of Assistant Professor or equivalent positions in Universities/Colleges / Institutions subject to the fulfilment of the following conditions: -

a) The Ph.D. degree of the candidate has been awarded in the regular mode

b) The Ph.D. thesis has been evaluated by at least two external examiners;

c) Open Ph.D. viva voce of the candidate has been conducted;

d) The candidate has published two research papers from his/her Ph.D. work out of which at least one is in a refereed journal;

e) The candidate has presented at least two papers based on his/her Ph.D. work in conferences/seminars sponsored /funded/supported by the UGC/ICSSR/CSIR or any similar agency.

Note: (i)The fulfillment of these conditions is to be certified by the Registrar or the Dean (Academic Affairs) of the University concerned. (ii)NET/SLET/SET shall also not be required for candidates in such Master's Programmes for which NET/SLET/SET is not conducted by the UGC, CSIR or similar test accredited by the UGC like SLET/SET.

**II. University Deputy Librarian**

i) A Master's Degree in Library Science/Information Science/ Documentation Science, with at least 55% marks or an equivalent grade in a point–scale, wherever grading system is followed.

ii) Eight years' experience as an Assistant University Librarian/College Librarian.

iii) Evidence of innovative library services, including integration of ICT in library.

iv) A Ph.D. Degree in Library Science/ Information Science / Documentation Science/Archives and manuscript keeping/computerization of library.

**III. University Librarian**

i) A Master's Degree in Library Science/Information Science/ Documentation Science with at least 55% marks or an equivalent grade in a point-scale wherever the grading system is followed.

ii) At least ten years as a Librarian at any level in University Library or ten years of teaching as Assistant/Associate Professor in Library Science or ten years' experience as a College Librarian.

iii) Evidence of innovative library services, including the integration of ICT in a library.

iv) A Ph.D. Degree in Library Science/Information Science/Documentation /Archives and manuscript keeping.

**Recruitment of LIS professionals in the universities of Kerala.**

LIS professionals have a vital role in fulfilling the objectives of the libraries and their parent organization. In Kerala, the qualification of a specific category of library professionals is on par with that of the academic staff. The qualifications and experience should fulfill the norms prescribed by the UGC standard. It guaranteed a professional approach to delivering services to the library's clientele. The statute of the university concerned determines the qualification of the LIS professionals in the universities.

Kerala Government has unified qualifications, age, method of appointment, and pay scale for the post of University Librarian (UGC post) and Professional Assistant Grade-II (Library) in the universities of Kerala. It was ordered as per

Order No. G.O.(P) No. 37/2020/ H.ED., Dated 06-11-2020 from the Higher Education (B) Department, Government of Kerala, Thiruvananthapuram. As per the government order, University Librarian and Professional Assistant Grade-II (Library) recruitment details are specified below.

1. Name of Post: University Librarian (UGC post)

Department: Universities in Kerala

Method of appointment: Direct Recruitment (As per UGC norms)

Scale of Pay: UGC Scale

**Basic Qualifications**

1. A Master's Degree in Library Science/Information Science/ Documentation Science with at least 55 % marks or an equivalent grade in a point-scale wherever the grading system is followed.
2. At least ten years as Librarian at any level -Assistant Librarian/Deputy Librarian (as per feeder Category) or equivalent in University Library or ten years of teaching as Assistant/ Associate Professor in Library Science or ten years' experience as a College Librarian
3. Evidence of innovative library services, including the integration of ICT in a library.
4. A Ph. D. Degree in Library Science/Information Science/Documentation/ Archives and Manuscript keeping.

Age limit: As per UGC norms

Appointing Authority: Vice-Chancellor

Probation: - Every person appointed to this post shall, from the date on which he joins duty, be on probation for a total period of two years on duty within a continuous period of three years. A probationer shall pass the prescribed departmental examination relating to probation.

1. Name of Post: Professionals Assistant Grade-II (Library)

Department: Universities in Kerala

Method of appointment: Direct Recruitment

Scale of Pay: Rs.27800-59400(pre-revised)

Basic Qualifications: MLISc in Library Science or BLISc or other equivalent qualification

Age limit: 22-36 years (with usual relaxation to Scheduled Castes, Scheduled Tribes and Other Backward Communities).

Appointing Authority: Registrar

Probation: From the date on which he joins duty, every person appointed to this post shall be on probation for a total period of two years on duty within a continuous period of three years. A probationer shall pass the prescribed departmental examination relating to probation.

## Conclusion

The chapter contains the profile of the selected university libraries in Kerala. The various aspects related to the study area are presented in this chapter. A library with a proper facility would give a sense of belonging, hope, and confidence to face challenges. The government sees fund allocation of libraries as an investment. It is pointed out that the Librarian post in most universities has been vacant for a long time. It may affect enhanced library facilities. More attention is required for the appointment of a librarian in the universities to make them vibrant academic institutions. All the stakeholders of the universities must come forward to improve the university library.

## References

Annual Report 2019 (2020). University of Calicut.

Annual Report 2019 (2020). University of Kerala. Thiruvananthapuram: University of Kerala.

Economic Review,2020(2021). Thiruvananthapuram: Kerala State Planning Board. Vol. I.

https://covas-opac.l2c2.co.in/cgi-bin

https://hgcl.kannuruniversity.ac.in/Pages/index.html

https://library.ssus.ac.in/index.php

https://libraryscience.uoc.ac.in/

https://mguniversity.ac.in

https://uoc.ac.in

https://www.cusat.ac.in.

https://www.kannuruniversity.ac.in/en/academics/

https://www.kvasu.ac.in/

India (2021). Report of the Comptroller and Auditor General of India on Economic Sector for the year ended March 2019. Government of Kerala Report No. 4 of the year 2021.

Kerala (2021). Report of the 11th Pay Revision Commission Government of Kerala., Part III.

UGC (2018). UGC Regulation 2018 notified in the Gazette of India, Part III, Section 4, July 18, 2018. Available at https://www.ugc.ac.in/ugc_regulations.aspx

# 6

# Data Analysis and Interpretation

This chapter presents the analysis and interpretation of data gathered through a structured questionnaire to conduct the current study. Data analysis and interpretation of data is the most significant part of the research report. It contains a detailed analysis of major results about the information for arriving at a valid conclusion. It helps to interpret the results of the study in a proper perspective. Based on the objectives and hypotheses formulated, the data was collected through a questionnaire. Likert's scale was used to assess the opinions of the respondents. The collected data were tabulated, and various statistical analyses were done to formulate the result of the study.

The Quality of Work Life (QWL) in the libraries of the universities of Kerala is analyzed. An extensive analysis was also conducted to find out the extent of job enlargement and job enrichment in the university libraries. This chapter also investigates and presents perceptions of Library and Information Science (LIS) professionals toward users and services.

In the present study, university libraries refer to university libraries/central libraries, departmental libraries, study center libraries, and constituent college libraries established and maintained to use their academic community by universities in Kerala. Hence, the present study is limited to university libraries/central libraries, departmental libraries, study center libraries, and constituent college libraries of eight prominent state universities in Kerala. The libraries of the universities selected for the study are Kerala University Library (KUL), University of Calicut Library (UoCL), Mahatma Gandhi University, Kerala Library (MGUL), Cochin University of Science and Technology Library (CUSATL), Kannur University Library (KAUL), Sree Sankaracharya University of Sanskrit Library (SSUSL), Kerala Agricultural University Library and Information System (KAULIS) and Kerala Veterinary and Animal Science University Library (KVASUL). Hence, the data presented in this chapter are responses from the eight selected state university libraries in Kerala.

The study is divided into the following broad areas

1. Profile of the study
2. Quality of Work Life in University Libraries
3. Job Enrichment in Libraries
4. Job Enlargement in Libraries
5. Perception of the LIS professionals towards the users and services

## I. Profile of the Study

The study was carried out among the two hundred and sixty-six LIS professionals. The study data was based on the responses collected from permanent and qualified two hundred and sixty-six professionally qualified LIS professionals working in the eight-state universities in Kerala. The study investigated background factors such as the name of the universities, the cadre of the LIS professionals, and characteristics of the professionals such as age, gender, and professional qualifications of the LIS professionals.

### 1.1 University-wise Distribution of Respondents

Table 6.1 presents the university-wise distribution of respondents and the names of the universities selected for the study, frequency, and percentage of respondents to the study.

**Table 6.1:** University-wise Distribution of Respondents

| Sl.No. | Name of Universities | Frequency | Per cent |
|---|---|---|---|
| 1. | University of Kerala (UoK), Thiruvananthapuram | 75 | 28.20 |
| 2. | University of Calicut (UoC), Thenjippalam | 60 | 22.56 |
| 3. | Mahatma Gandhi University, Kerala (MGUK), Kottayam | 42 | 15.79 |
| 4. | Cochin University of Science and Technology (CUSAT), Kochi | 35 | 13.16 |
| 5. | Kannur University (KU), Kannur | 23 | 8.65 |
| 6. | Sree Sankaracharya University of Sanskrit (SSUS), Kalady | 15 | 5.64 |
| 7. | Kerala Agricultural University (KAU),Vellanikkara | 10 | 3.76 |
| 8. | Kerala Veterinary and Animal Sciences University (KVASU), Pookot | 6 | 2.26 |
| | Total | 266 | 100.00 |

Table 6.1 demonstrates that 75 (28.20 percent) respondents were from the University of Kerala, and 60 (22.56 percent) respondents were from the University of Calicut. Respondents selected from Mahatma Gandhi University,

Kerala, are 42 (15.79 percent), and those chosen from Cochin University of Science and Technology are 35 (13.16 percent). The total number of respondents from Kannur University is 23 (8.65 percent). Respondents from Sree Sankaracharya University of Sanskrit are 15 (5.64 percent), and 10 (3.76 percent) respondents from Kerala Agricultural University. Respondents from Kerala Veterinary and Animal Sciences University are 6 (2.26 percent). The table reveals that 266 LIS professionals are working in Kerala's selected eight state universities. The study showed that most respondents are from the University of Kerala, followed by the University of Calicut. The lowest number of respondents to the study are from the Kerala Veterinary and Animal Science University.

## 1.2 Cadre-wise Distribution of Respondents

Human resources have a vital role in the library. Qualified and competent manpower plays a prominent role in successfully managing libraries and information centers. There are different cadres among the Library and Information (LIS) professionals. The entry-level cadre of LIS professionals is Library Assistants or Professional Assistants Grade-II in the universities of Kerala. From the entry level, they are subsequently promoted to higher-level posts. In the post of Assistant Librarian and Deputy Librarian, both UGC/ICAR scale of pay and state scale of pay in the universities. University Librarian heads university Library. In Kerala Agricultural University and Kannur University, there is a Central Library which Librarian heads. The post of University Librarian or Librarian is a statutory post. There are seven cadres of LIS professionals by their cadre.

Table 6.2 shows a detailed cadre-wise distribution of LIS professionals in Kerala's selected state university libraries.

**Table 6.2:** Cadre-wise Distribution of Respondents

| Cadre of LIS Professionals | | Name of the University Libraries | | | | | | | | Total |
|---|---|---|---|---|---|---|---|---|---|---|
| | | KUL | UOCL | MGUL | CUSATL | KAUL | SSUSL | KAULIS | KVASUL | |
| University Librarian | n | 0 | 0 | 0 | 0 | 0 | 0 | 0 | 0 | 0 |
| Deputy Librarian | n | 3 | 1 | 3 | 3 | 0 | 1 | 1 | 0 | 12 |
| | % | 4.00 | 1.67 | 7.14 | 8.57 | 0.00 | 6.67 | 10.00 | 0.00 | 4.51 |
| Assistant Librarian | n | 31 | 16 | 7 | 16 | 1 | 0 | 4 | 2 | 77 |
| | % | 41.33 | 26.67 | 16.67 | 45.71 | 4.35 | 0.00 | 40.00 | 33.33 | 28.95 |
| Reference Assistant HG/ Reference Officer (HG) | n | 0 | 0 | 0 | 0 | 0 | 0 | 4 | 1 | 5 |
| | % | 0.00 | 0.00 | 0.00 | 0.00 | 0.00 | 0.00 | 40.00 | 16.67 | 1.88 |

| Cadre of LIS Professionals | | Name of the University Libraries | | | | | | | | Total |
|---|---|---|---|---|---|---|---|---|---|---|
| | | KUL | UOCL | MGUL | CUSATL | KAUL | SSUSL | KAULIS | KVASUL | |
| Reference Assistant/ Junior Librarian/ Reference Officer | n | 14 | 13 | 10 | 8 | 5 | 2 | 0 | 2 | 54 |
| | % | 18.67 | 21.67 | 23.81 | 22.86 | 21.74 | 13.33 | 0.00 | 33.33 | 20.30 |
| Technical Assistant/ Professional Assistant Gr. I | n | 25 | 22 | 15 | 7 | 10 | 5 | 0 | 0 | 84 |
| | % | 33.33 | 36.67 | 35.71 | 20.00 | 43.48 | 33.33 | 0.00 | 0.00 | 31.58 |
| Professional Assistant Gr-II/ Library Assistant | n | 2 | 8 | 7 | 1 | 7 | 7 | 1 | 1 | 34 |
| | % | 2.67 | 13.33 | 16.67 | 2.86 | 30.43 | 46.67 | 10.00 | 16.67 | 12.78 |
| Total | n | 75 | 60 | 42 | 35 | 23 | 15 | 10 | 6 | 266 |
| | % | 100 | 100 | 100 | 100 | 100 | 100 | 100 | 100 | 100 |

Table 6.2 indicates the cadre-wise distribution of the LIS professionals in the universities. The University Librarian cadre is the head of the university library lying vacant in all the universities. Next to the University Librarian is the Deputy Librarian cadre, and respondents working in the cadre of Deputy Librarians are 12 (4.51 percent), and Assistant Librarians are 77 (28.95 percent). The total number of Reference Assistants (Higher Grade)/Reference Officers (Higher Grade) is 5 (1.88 percent), and Reference Assistant/Junior Librarian/ Reference Officer is 54 (20.30 percent). Respondents are working in the cadre of Technical Assistant/Professional Assistant Grade. I am 84 (31.58 percent) and 34 (12.78 percent) respondents working in the Professional Assistant Gr-II/Library Assistant cadre. The study shows that a maximum number of respondents are working in the cadre of Assistant Librarians, followed by Technical Assistants/Professional Assistants Grade.I in the universities of Kerala. The study revealed that more than 50 percent of respondents work in supervisory and above-level positions in the universities of Kerala.

### 1.3 Personal Characteristics

Personal characteristics considered for the study are the age and gender of respondents.

#### 1.3.1 Age of Respondents

Age was operationally defined as the number of years completed by the respondent in the current investigation. Age is a characteristic that helps to find out the potential population in the organization. The respondents of the study, based on their age, are classified into four groups. The groups are up to or below 40 years, 41 to 45, 46 to 50, and above 50.

Table 6.3 shows the age-wise distribution of respondents.

**Table 6.3: Age-wise Distribution of Respondents**

| Sl.No. | Age group (Years) | Frequency | Per cent |
|---|---|---|---|
| 1 | Up to or below 40 | 87 | 32.7 |
| 2 | 41-45 | 61 | 22.9 |
| 3 | 46-50 | 61 | 22.9 |
| 3 | Above 50 | 57 | 21.4 |
| | Total | 266 | 100 |

Table 6.3 indicates that 87 (32.7 percent) respondents are in the age group of up to 40 years or below 40 years of age, and 61 (22.9 percent) respondents are in the age group of 41-45 years. Respondents in the age group of 46-50 years are 61 (22.9 percent), and those in the age group above 50 are 57 (21.4 percent). It may be inferred from the study that the majority of the library staff in the universities of Kerala are below or up to 40 years of age. It indicates that there is scope for improving their job skills.

### 1.1.2 Gender of Respondents

Both male and female LIS professionals working in the university are selected as the current study respondents. Gender is a socio-demographic variable. It is used to find out the gender status. Gender is an independent variable. It is used to test with other dependent variables selected for the study.

Table 6.4 presents the gender-wise distribution of respondents.

**Table 6.4:** Gender–Gender-wise distribution of Respondents

| Sl.No. | Gender | Frequency | Per cent |
|---|---|---|---|
| 1 | Male | 118 | 44.4 |
| 2 | Female | 148 | 55.6 |
| | Total | 266 | 100 |

Table 6.4 shows that the number of female respondents in the universities is greater than that of male respondents. The percentage of male respondents was 44.4 percent, and female respondents were 55.6 percent. Onuoha., Samuel, and Ojo (2014) opined that library administrators adopt a policy of equal treatment for each gender and review the conditions attached to promotion, among others, to enhance job satisfaction.

### 1.4 Professional Qualifications of Respondents

Table 6.5 presents the Professional qualification distribution of respondents in the universities of Kerala.

**Table 6.5:** Professional Qualification wise distribution of Respondents

| Sl.No. | **Professional Qualifications** | **n** | **Per cent** |
|---|---|---|---|
| 1 | Bachelor of Library and Information Science (BLISc) | 17 | 6.39 |
| 2 | Master of Library and Information Science (MLISc) | 161 | 60.53 |
| 3 | Master of Philosophy (M. Phil.) | 49 | 18.42 |
| 4 | Doctor of Philosophy (Ph. D.) | 39 | 14.66 |
| | Total | 266 | 100 |

Table 6.5 indicates that 17 (6.39 percent) respondents have a BLISc degree, and 161 (60.53 percent) respondents have an MLISc Degree. Respondents with an M. Phil degree in LIS are 49(18.42 percent), and 39 (14.66 percent) respondents have a Ph.D. in LIS. It shows that only 6.39 percent of respondents have entry-level qualifications, and others have higher-level qualifications in LIS.

## 2. Quality of Work Life in the University Libraries

The success of an organization mainly depends on how the organization attracts, motivates, and retains its personnel. Quality of work life (QWL) is what an employee experiences at his workplace. QWL is vital in shaping attitudes towards work life and relationships with clientele and authorities. It will encourage constitutionalism of work, opportunities for continuous growth and security, and safe and healthy working conditions in the library. It affects employee commitment, engagement, job involvement, and the reputation of the university library. Organizational climate and job motivation influence the QWL of library and information professionals working in the university libraries in Kerala. The study shows that LIS professionals belonging to different levels of organizational climate have different dimensions of quality of work life.

The quality of work life of library and Information professionals is depicted in Table 6.6.

**Table 6.6:** Quality of Work Life of Library and Information Professionals

| Statements | | Strongly Agree | Agree | Neither Agree Nor Disagree | Disagree | Strongly Disagree | Mean | SD | t | Sig. |
|---|---|---|---|---|---|---|---|---|---|---|
| Inadequate and fair compensation exists for staff | n | 10 | 77 | 109 | 67 | 3 | 3.09 | 0.86 | -1.721 | 0.086 |
| | % | 3.76 | 28.95 | 40.98 | 25.19 | 1.13 | | | | |
| There exists a safe and healthy working condition | n | 14 | 171 | 47 | 30 | 4 | 3.61 | 0.81 | -12.128 | 0.000 |
| | % | 5.26 | 64.29 | 17.67 | 11.28 | 1.50 | | | | |
| No Immediate opportunities to use and develop human capacity | n | 7 | 82 | 100 | 74 | 3 | 3.06 | 0.86 | -1.144 | 0.254 |
| | % | 2.63 | 30.83 | 37.59 | 27.82 | 1.13 | | | | |
| Opportunities for continues growth and security is inadequate | n | 8 | 80 | 86 | 89 | 3 | 3.00 | 0.90 | -0.068 | 0.946 |
| | % | 3.01 | 30.08 | 32.33 | 33.46 | 1.13 | | | | |
| Opportunities for social integration are not ample. | n | 7 | 60 | 104 | 94 | 1 | 2.92 | 0.83 | 1.618 | 0.107 |
| | % | 2.63 | 22.56 | 39.10 | 35.34 | 0.38 | | | | |
| There is constitutionalism in the work | n | 2 | 85 | 142 | 36 | 1 | 3.19 | 0.68 | -4.586 | 0.000 |
| | % | 0.75 | 31.95 | 53.38 | 13.53 | 0.38 | | | | |
| Social relevance of equality of rules, work assignment etc are prevailing in the library. | n | 8 | 101 | 103 | 50 | 4 | 3.22 | 0.84 | -4.319 | 0.000 |
| | % | 3.01 | 37.97 | 38.72 | 18.80 | 1.50 | | | | |
| Entire work in the library is not favourably related with my total life space. | n | 3 | 45 | 110 | 97 | 11 | 2.74 | 0.83 | 5.050 | 0.000 |
| | % | 1.13 | 16.92 | 41.35 | 36.47 | 4.14 | | | | |

From Table 6.6, it is clear that in the quality of work-life out of the total respondents, 3.76 percent strongly agree and 28.95 percent agree. Nearly 40.98 percent neither agree nor disagree. Only 25.19 percent of respondents agree, and 1.13 percent strongly disagree. The mean score of the statement *Inadequate and fair compensation exists for staff* is 3.09, which is significantly higher than the means of the response scale as the significance level of the t-test is less than 0.05. It means that inadequate and fair compensation exists for library staff. The opinion score of the statement *There exists a safe and healthy working condition* is found to be the highest score of 3.61, which indicates a safe and healthy working condition in the library. The second important aspect of quality of work-life is *the Social relevance of equality of rules, work assignments, etc., prevailing in the library.* The mean score of that aspect is 3.22. The other important aspects of quality of work-life in order of descending level are *There is constitutionalism in work* (3.19*), No Immediate opportunities to use and develop human capacity* (3.06), and *Opportunities for continuous growth and security are inadequate* (3.00). The mean score of these aspects is significantly higher than the mean of the response scale. The other important aspect of quality of work life is *that opportunities for social integration are insufficient* (2.92). It is significantly lower than the mean of the response scale. The lowest mean score found for *the Entire work in the library is not favorably related to my total life space* (2.74). It means that the entire work in the library is not favorably related to library professionals' total life space.

Table 6.7 depicts the mean opinion score of library and Information professionals about QWL among Universities in Kerala.

**Table 6.7:** Mean Opinion Score of Library and Information Professionals about QWL

| Statements | Statistics | Name of Universities | | | | | | | | ANOVA | | MANOVA | |
|---|---|---|---|---|---|---|---|---|---|---|---|---|---|
| | | KUL | UoCL | MGUL | CUSATL | Kaul | SSUSL | KAULIS | KVASUL | F | Sig. | F | |
| Inadequate and fair compensation exists for staff | Mean | 3.23 | 3.03 | 2.98 | 2.74 | 3.09 | 3.20 | 3.80 | 3.33 | 2.418 | 0.021 | 2.426 | 0.000 |
| | SD | 0.91 | 0.80 | 0.75 | 0.89 | 0.85 | 0.86 | 0.42 | 1.03 | | | | |
| There exists a safe and healthy working condition | Mean | 3.65 | 3.77 | 3.50 | 3.71 | 2.74 | 3.67 | 3.90 | 4.17 | 5.470 | 0.000 | | |
| | SD | 0.83 | 0.56 | 0.74 | 0.79 | 1.10 | 0.72 | 0.57 | 0.75 | | | | |
| No immediate opportunities to use and develop human capacity | Mean | 2.97 | 3.17 | 3.10 | 2.91 | 3.39 | 3.13 | 2.30 | 3.50 | 2.327 | 0.026 | | |
| | SD | 0.84 | 0.89 | 0.82 | 0.78 | 0.89 | 0.92 | 0.48 | 1.05 | | | | |
| Opportunities for continues growth and security is inadequate | Mean | 3.09 | 2.92 | 2.93 | 2.83 | 3.70 | 2.80 | 2.60 | 2.83 | 2.954 | 0.005 | | |
| | SD | 0.93 | 0.87 | 0.78 | 0.92 | 0.76 | 0.86 | 0.97 | 0.75 | | | | |
| Opportunities for social integration are not ample. | Mean | 3.00 | 2.73 | 2.90 | 2.91 | 3.52 | 2.73 | 2.70 | 2.33 | 3.027 | 0.004 | | |
| | SD | 0.85 | 0.80 | 0.79 | 0.85 | 0.79 | 0.80 | 0.67 | 0.52 | | | | |
| There is constitutionalism in the work | Mean | 3.16 | 3.40 | 3.10 | 2.97 | 3.04 | 3.40 | 3.30 | 3.33 | 1.940 | 0.064 | | |
| | SD | 0.77 | 0.67 | 0.58 | 0.57 | 0.71 | 0.63 | 0.67 | 0.52 | | | | |
| Social relevance of equality of rules, work assignment etc are prevailing in the library. | Mean | 3.09 | 3.45 | 3.10 | 3.20 | 3.13 | 3.00 | 3.60 | 3.83 | 2.019 | 0.053 | | |
| | SD | 0.92 | 0.70 | 0.79 | 0.87 | 0.92 | 0.85 | 0.52 | 0.75 | | | | |
| Entire work in the library is not favourably related with my total life space. | Mean | 2.79 | 2.48 | 2.86 | 2.49 | 3.22 | 3.27 | 2.90 | 2.17 | 4.227 | 0.000 | | |
| | SD | 0.84 | 0.81 | 0.61 | 0.82 | 0.85 | 0.88 | 0.57 | 0.75 | | | | |

Table 6.7 reveals that in the matter of QWL, inadequate and fair compensation exists for staff in the job of LIS professionals, which varies among the universities of Kerala. It is found to be highest in the KAULIS, followed by KUL, and lowest in CUSATL.

Professionals with safe and healthy working conditions vary significantly among the different university libraries in Kerala, as the significance level of ANOVA is less than 0.05. It is found to be highest in KVASUL, with a mean score of 4.17, followed by KAULIS, with a mean score of 3.90. It is found to be the lowest in KAUL, with a mean score of 2.74.

The statement "No immediate opportunities to use and develop human capacity" is found to be highest in the CUSATL, with a mean score of 3.50, followed by KAUL, with a mean score of 3.39. It is found to be the lowest in the KAULIS, with a mean score of 2.30. LIS professionals with "Opportunities for continued growth and security is inadequate" are found to be highest in KUL with a mean score of 3.09, followed by MGUL. It is found to be lowest in KAULIS, with a mean score of 2.60. The statement "Opportunities for social integration are not ample" is highest in KAUL, with a mean score of 3.52, and it is found to be lowest in KVASUL, with a mean score of 2.33.

Regarding the constitutionalism of work of LIS professionals, it is found to be highest in UOCL and SSUL with a mean score of 3.40, and it is found to be lowest in CUSATL with a mean score of 2.97. The statement "Social relevance of equality of rules, work assignment, etc., are prevailing in the library" is highest in KVASUL with a mean score of 3.83, followed by KAULIS with a mean score of 3.60. LIS professionals with the Entire work in the library are not favourably related to my total life space, which varies significantly among the different universities in Kerala as the significance level of ANOVA is less than 0.05. It is highest in SSUSL, with a mean score of 3.27, followed by KAUL, with a mean score of 3.22. It is found to be lowest in KVASUL, with a mean score of 2.17.

Figure 6.1 illustrates the mean opinion score of library and Information professionals about QWL.

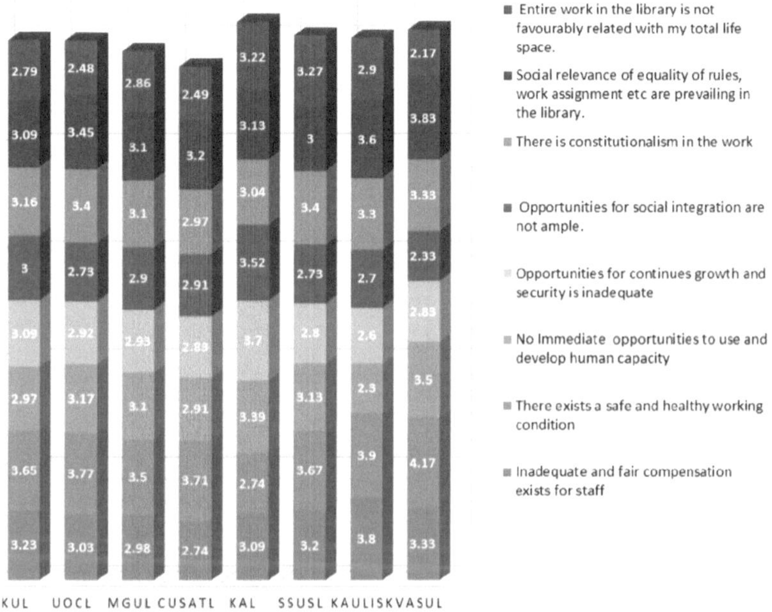


**Fig. 6.1:** Mean Opinion Score of Library and Information Professionals about QWL

There is a significant difference in the aspects of Quality of Work Life (QWL) as the significance level of ANOVA is greater than 0.05. However, when all aspects of QWL are taken together, there is a significant variation concerning libraries of universities in Kerala, as the significance level of MANOVA is less than 0.05. The study shows that the QWL of the libraries of universities in Kerala is different. A significant difference exists in QWL among LIS professionals working in the universities in Kerala. Darmawansyah et al. (2018) explained that the quality of work-life and organizational climate positively and significantly affects work satisfaction.

## 3. Job Enrichment in Libraries

Job enrichment assigns responsibilities usually reserved for a higher level of employees. Job enrichment allows complete autonomy to plan and carry out all activities on the job. In a library, job enrichment enables LIS professionals to learn new skills. It creates an opportunity for the advancement and effective dissemination of information in an organization.

Table 6.8 presents job enrichment among LIS professionals in the selected state universities in Kerala.

**Table 6.8:** Job Enrichment among LIS Professionals

| Statements | | Strongly Agree | Agree | Neither Agree Nor Disagree | Disagree | Strongly Disagree | Mean | SD | t | Sig. |
|---|---|---|---|---|---|---|---|---|---|---|
| I am assigned responsibilities normally reserved for higher level employees. | n | 7 | 75 | 105 | 77 | 2 | 3.03 | 0.84 | -0.583 | 0.561 |
| | % | 2.63 | 28.20 | 39.47 | 28.95 | 0.75 | | | | |
| Not given complete autonomy to plan and carry out my activities on the job. | n | 2 | 76 | 95 | 88 | 5 | 2.93 | 0.85 | 1.301 | 0.194 |
| | % | 0.75 | 28.57 | 35.71 | 33.08 | 1.88 | | | | |
| I require high knowledge and skill to perform my job. | n | 8 | 114 | 85 | 58 | 1 | 3.26 | 0.85 | -5.074 | 0.000 |
| | % | 3.01 | 42.86 | 31.95 | 21.80 | 0.38 | | | | |
| Our library allows me to learn new skills. | n | 7 | 146 | 78 | 31 | 4 | 3.46 | 0.79 | -9.375 | 0.000 |
| | % | 2.63 | 54.89 | 29.32 | 11.65 | 1.50 | | | | |
| Performing additional duty is not connected with offering additional financial benefits. | n | 9 | 148 | 82 | 24 | 3 | 3.51 | 0.75 | -11.066 | 0.000 |
| | % | 3.38 | 55.64 | 30.83 | 9.02 | 1.13 | | | | |
| Motivation is important to do the vertically loaded jobs | n | 15 | 151 | 89 | 10 | 1 | 3.64 | 0.67 | -15.553 | 0.000 |
| | % | 5.64 | 56.77 | 33.46 | 3.76 | 0.38 | | | | |
| Opportunity is not given in our library to use my variety of skills. | n | 6 | 66 | 104 | 85 | 5 | 2.94 | 0.86 | 1.219 | 0.224 |
| | % | 2.26 | 24.81 | 39.10 | 31.95 | 1.88 | | | | |
| There are no sufficient opportunities for advancement in our organization. | n | 11 | 80 | 86 | 80 | 9 | 3.02 | 0.95 | -0.258 | 0.797 |
| | % | 4.14 | 30.08 | 32.33 | 30.08 | 3.38 | | | | |

Table 6.8 indicates that out of the total respondents, 2.63 percent strongly agree, and 28.20 percent agree. Nearly 39.47 percent of respondents neither agree nor disagree. Only 28.95 percent of the respondents disagree, and 0.75 percent of respondents strongly disagree. The mean score of the statement *I am assigned responsibilities normally reserved for higher-level employees* is 3.03, which is significantly greater than the mean of the response scale as the significance level of the t-test is lower than 0.05. It indicates that there are instances of assigning responsibilities normally reserved for higher levels of employees. The opinion score of the statement *Motivation is important to do the vertically loaded jobs* is found to be the highest score of 3.64, which is significantly greater than the mean of the response scale. It indicates that motivation is important to do vertically loaded jobs. The important aspect of job enrichment is *that performing additional duties is not connected with offering additional financial benefits.* The mean score of that aspect is 3.51. The other important elements of job enrichment in order of descending level are: *Our library allows me to learn new skills* (3.46), *I require high knowledge and skill to perform my job* (3.26), and *there are no sufficient opportunities for advancement in our organization* (3.02). The other important aspect of job enrichment is *that the opportunity to use my variety of skills* is not given in our library (2.94). The lowest score is found for *Not being given complete autonomy to plan and carry out my activities on the job* (2.93). The result indicates that library professionals are not given full autonomy to plan and carry out their activities.

To find a composite variable representing the level of job enrichment, the average score of its dimensions was computed, and the descriptive statistics of this variable are given below. The composite variable's mean score is 3.22, with a standard deviation of 0.36. The median is 3.13, which is a little smaller than the mean. The score of the variable ranges from a minimum of 2.38 to a maximum of 5.00. Table 6.9 depicts descriptive statistics of the job enrichment score.

**Table 6.9:** Descriptive Statistics of Score of Job Enrichment

| Statistics | Values | t | Sig. |
|---|---|---|---|
| Mean | 3.22 | 9.944 | 0.000 |
| Median | 3.13 | | |
| SD | 0.36 | | |
| Variance | 0.13 | | |
| Minimum | 2.38 | | |
| Maximum | 5.00 | | |

Table 6.9, the one-sample t-test, indicates that the mean of the variable is significantly higher than the mean of the response scale. There is a significantly high job enrichment level among the library and information professionals of university libraries in Kerala.

Table 6.10 presents a variation of the job enrichment score among libraries of the universities selected for the study.

**Table 6.10:** Variations of Score of Job Enrichment among Libraries of Universities

| Name of University Libraries | Mean | SD | ANOVA | |
|---|---|---|---|---|
| | | | F | Sig. |
| Kerala University Library (KUL) | 3.17 | 0.37 | 1.740 | 0.100 |
| University of Calicut Library (UoCL) | 3.26 | 0.39 | | |
| Mahatma Gandhi University, Kerala Library (MGUL) | 3.26 | 0.41 | | |
| Cochin University of Science and Technology Library (CUSATL) | 3.11 | 0.30 | | |
| Kannur University Library (KAUL) | 3.30 | 0.34 | | |
| Sree Sankaracharya University of Sanskrit Library (SSUSL) | 3.41 | 0.37 | | |
| Kerala Agricultural University Library and Information System (KAULIS) | 3.13 | 0.18 | | |
| Kerala Veterinary and Animal Science University Library (KVASUL) | 3.15 | 0.15 | | |
| Total | 3.22 | 0.36 | | |

In Table 6.10, the variation of the score of job enrichment among the libraries of universities is assessed by one way ANOVA. The result of the ANOVA shows no significant variation in the level of job enrichment among the university libraries, as the significance level is greater than 0.05. From the study, it can be inferred that as far as job enrichment is concerned, the situation of all the university libraries in Kerala is high and alike.

### 4. Job Enlargement in Libraries

Job enlargement is the horizontal loading of a job. It creates the opportunity to perform various jobs at the same level in the library. It assigns employees tasks other than regular duties and responsibilities and helps improve their abilities, talent caliber, and skills.

Table 6.11 presents job enlargement among library and information professionals.

**Table 6.11:** Job Enlargement among Library and Information Professionals

| Statements | | Strongly Agree | Agree | Neither Agree Nor Disagree | Disagree | Strongly Disagree | Mean | SD | t | Sig. |
|---|---|---|---|---|---|---|---|---|---|---|
| I shall frequently transfer from one section to another | n | 8 | 37 | 81 | 127 | 13 | 2.62 | 0.89 | 6.881 | 0.000 |
| | % | 3.01 | 13.91 | 30.45 | 47.74 | 4.89 | | | | |
| No scope of increasing the experience of the job | n | 3 | 46 | 105 | 106 | 6 | 2.75 | 0.80 | 5.028 | 0.000 |
| | % | 1.13 | 17.29 | 39.47 | 39.85 | 2.26 | | | | |
| Not satisfied with the type of professional experience which is getting in our library. | n | 8 | 65 | 80 | 106 | 7 | 2.85 | 0.92 | 2.593 | 0.010 |
| | % | 3.01 | 24.44 | 30.08 | 39.85 | 2.63 | | | | |
| I have opportunities to perform various jobs at the same level in the library | n | 7 | 133 | 90 | 35 | 1 | 3.41 | 0.76 | -8.833 | 0.000 |
| | % | 2.63 | 50.00 | 33.83 | 13.16 | 0.38 | | | | |
| Performing additional duties does not satisfy my needs of self-esteem, autonomy etc | n | 2 | 42 | 115 | 102 | 5 | 2.75 | 0.77 | 5.281 | 0.000 |
| | % | 0.75 | 15.79 | 43.23 | 38.35 | 1.88 | | | | |
| To assign employees additional tasks other than their regular job duties. | n | 2 | 77 | 114 | 69 | 4 | 3.02 | 0.80 | -0.306 | 0.760 |
| | % | 0.75 | 28.95 | 42.86 | 25.94 | 1.50 | | | | |
| I perform the additional tasks as efficiently as the regular tasks. | n | 9 | 137 | 101 | 19 | 0 | 3.51 | 0.68 | -12.266 | 0.000 |
| | % | 3.38 | 51.50 | 37.97 | 7.14 | 0.00 | | | | |
| Assigning me with additional tasks does not improve my abilities and skills. | n | 2 | 33 | 98 | 117 | 16 | 2.58 | 0.81 | 8.454 | 0.000 |
| | % | 0.75 | 12.41 | 36.84 | 43.98 | 6.02 | | | | |

Table 6.11 reveals that out of the total respondents, 3.01 percent strongly agree, and 13.91 percent agree. Nearly 30.45 percent of respondents neither agree nor disagree. Only 47.74 percent disagree, and 4.89 percent strongly agree. The mean score of the statement *I shall frequently transfer from one section to another* is found to be the lowest score of 2.62, which is significantly lower than the mean of the response scale as the significance level of the t-test is greater than 0.05. It indicates that library professionals disagree that they are frequently transferred from one section to another section. The score of the statement that *I perform the additional tasks as efficiently as the regular tasks* is the highest score of 3.51, which indicates that library professionals perform the additional tasks as efficiently as the regular tasks. The second important aspect of job enlargement is *my opportunities to perform various jobs at the same level in the library* (3.41). It is significantly higher than the mean of the response scale. The other important aspect of job enlargement is "*To assign employees additional tasks other than their regular job duties* (3.02)". It is significantly higher than the mean of the response scale. The other important aspects of job enlargement in order of descending level are *Not satisfied with the type of professional experience which is getting in our library* (2.85), *No scope of increasing the experience of the job, performing additional duties does not satisfy my needs of self-esteem, autonomy, etc.* (2.75 each). The lowest score is found for *Assigning me additional tasks that do not improve my abilities and skills* (2.58). It indicates that library professionals have differing views on assigning them additional tasks, not improving their abilities and skills.

Table 6.12 presents descriptive statistics of the job enlargement score.

**Table 6.12:** Variation of Score of Job Enlargement among Libraries of Various Universities

| **Statistics** | **Values** | **One sample t-test for mean** | |
|---|---|---|---|
| | | **t** | **Sig.** |
| Mean | 2.94 | 2.618 | 0.009 |
| Median | 2.88 | | |
| SD | 0.39 | | |
| Variance | 0.15 | | |
| Minimum | 2.00 | | |
| Maximum | 4.88 | | |

Table 6.12 depicts that the one-sample t-test for the mean job enlargement score indicates that the variable's mean is significantly lower than the mean of the response scale. There is no significant level of job enlargement among library professionals.

Table 6.13 presents the variation of the job enlargement score among university libraries.

**Table 6.13:** Variation of Score of Job Enlargement among Libraries of Universities

| Name of University Libraries | Mean | SD | ANOVA | |
|---|---|---|---|---|
| | | | F | Sig. |
| Kerala University Library (KUL) | 2.99 | 0.38 | 3.559 | 0.001 |
| University of Calicut Library (UoCL) | 2.90 | 0.46 | | |
| Mahathma Gandhi University, Kerala Library (MGUL) | 2.89 | 0.39 | | |
| Cochin University of Science and Technology Library (CUSATL) | 2.82 | 0.29 | | |
| Kannur Univeristy Library (KAUL) | 3.05 | 0.27 | | |
| Sree Sankaracharya University of Sanskrit Library (SSUSL) | 3.27 | 0.38 | | |
| Kerala Agricultural University Library and Information System (KAULIS) | 2.90 | 0.22 | | |
| Kerala Veterinary and Animal Science University Library (KVASUL) | 2.58 | 0.23 | | |
| Total | 2.94 | 0.39 | | |

Table 6.13 shows that one-way ANOVA assessed the variation score of job enlargement among the various university libraries. The result of the ANOVA shows significant variation in job enlargement among various university libraries as the significance level is less than 0.05. As far as job enlargement is concerned, all libraries of universities in Kerala are low and different. The job enlargement of library and information professionals is low and different in the university libraries in Kerala. Job Enlargement is found to be highest in the SSUSL, followed by the KAUL, and it is found to be lowest in the KVASUL. Nandy (1985) opined that Job enrichment, job enlargement, and job rotation in the library field increase job satisfaction, and job satisfaction leads to the effective functioning of the individuals and the organization as a whole. Sharma Narendra Nath (2016) emphasized that libraries and institutes should have the courage to motivate LIS professionals through effective job analysis.

## 5. Perception of the library and information professionals towards the users and services

### Distribution of Libraries Based on the Perception of Users and Services

Perception about users and services is essential for examining an organization's Quality of life. Data was collected from the LIS professionals regarding their perception of users and services rendered by the libraries.

Table 6.14 presents the distribution of libraries based on users and services.

**Table 6.14:** Distribution of Libraries Based on the Perception of Users and Services

| Statement related to Service Quality | Yes | | No | |
|---|---|---|---|---|
| | Frequency | Per cent | Frequency | Per cent |
| Users think that the working hours of the library are sufficient for them. | 154 | 57.89 | 112 | 42.11 |
| We are giving need-based service to our clientele. | 244 | 91.73 | 22 | 8.27 |
| In our library clientele is highly satisfied with service | 202 | 75.94 | 64 | 24.06 |

Table 6.14 shows that 57.89 percent of respondents revealed that the library's working hours are sufficient for users, and 42.11 percent of respondents believe that the library's operating hours are not enough for users. About 91.73 percent of respondents reveal that they are giving need-based services to their clientele, and 8.27 percent believe that they are not providing need-based assistance. About 75.94 percent of the respondents reveal that the library clientele is delighted with the service of their library, and 24.06 percent of the respondents indicate that in their library, the clientele is not satisfied with their library service. It shows that LIS professionals are delivering services with satisfied clientele.

## Conclusion

The analysis and interpretation of data have identified the respondents' preference on the various aspects relating to the QWL of university libraries. From the study, the quality of work-life of the libraries of universities in Kerala is different. A significant difference exists in the quality of work-life among LIS professionals working in the universities in Kerala. There exists a significantly high level of job enrichment. However, there is no significant level of job enlargement among library professionals. The study also reveals that LIS professionals deliver services, and they are satisfied with their service. To improve the functioning of the libraries and information centers, policy should be formulated by giving importance to the factors mentioned above affecting Quality of Work Life.

## References

Darmawansyah, Rompu., Wahyu Atjo., & Abadi Muh. Yusri. (2018). The effect of occupational stress, Quality of Work life and Organizational Climate on Officials' Work Satisfaction of Regional Public Hospital of Undata Palu. Indian Journal of Public Health Research & Development,9(3),309-313. doi:10.5958/0976-5506.2018.00228.0

Nandy, Subodh Gopal. (1985). Job satisfaction of the library professionals. Herald of Library Science.24(4), 295-300.

Sharma, Narendra Nath. (2016). Job Analysis of Library Professionals. In Science and Technology Institutes of Assam:An Evaluative Study (Ph. D. Thesis), Gauhati University, Guwahati. Retrieved from http://hdl.handle.net/10603/144701

# 7

# Findings and Conclusion

## Introduction

This chapter gives an account of a summary of the significant findings of the study, followed by tenability of hypotheses, suggestions, further areas of research in the related areas, and a conclusion. The present study investigated the Quality of Work Life (QWL) of Library and Information Science (LIS) professionals working in Kerala's eight prominent state universities. The data of the study were collected from the two hundred and sixty-six 266 regular qualified LIS professionals. The data analysis and interpretation of the previous chapter revealed several valuable results. The concept of QWL in university libraries has been studied along with job enrichment, job enlargement, and perception of the library and information professionals towards the users and services. The result of the study was analyzed using MS Excel and SPSS, findings were formulated, and conclusions were finally reached.

## Findings of the Study

The following findings are derived based on the analysis and interpretation of data in the previous chapter.

- **To understand the Quality of Work Life in the university libraries of Kerala.**

The study was conducted among the eight state university libraries in Kerala. The study reveals that when all aspects of quality of Work Life are taken together, there exists a significant variation in the quality of work life of the libraries attached to the universities in Kerala.

- **To study the QWL among LIS professionals working in the University Libraries in Kerala.**

The study was conducted among the two hundred and sixty-six LIS professionals working in the university libraries of Kerala. The study revealed that professionals working in the different organizational climates of the selected university libraries have different perceptions about the quality of work life.

**• To compare the situation of QWL in the university libraries of Kerala.**

From Table 6.7 and Figure 6.1 of the previous chapter, the Quality of work-life varies significantly different among the university libraries in Kerala. But when all aspects relating to the Quality of Work Life are taken together, there is a significant variation in the libraries.

**• To study the extent of job enrichment in the university libraries of Kerala.**

Examination of Job enrichment among library and information science professionals shows that there exists a significantly high level of job enrichment among the library professionals working in the selected universities of Kerala. Moreover, regarding job enrichment, all university libraries in Kerala are high and alike. The highest score of job enrichment is found in the Sree Sankara University of Sanskrit Library (SSUSL), followed by the Kannur University Library (KAUL).

**• To study the extent of job enlargement in the university libraries of Kerala.**

The study indicates no significant level of job enlargement among the library and information professionals in the universities in Kerala. Regarding job enlargement, all libraries of the universities in Kerala are low and different. Among the selected universities, the highest score is found in Sree Sankara University of Sanskrit Library (SSUSL), followed by Kannur University Library (KAUL).

**• To investigate the perception of the library and information professionals. towards the users and services in the selected university libraries.**

It is proved from the data analysis that most of the LIS professionals agree that the library's working hours are sufficient. LIS professionals are giving need-based services to their clientele. The professionals agree that the library's clientele is delighted with the library's service. It reveals that LIS professionals are delivering services with satisfied clientele.

## Tenability of Hypotheses

The two hypotheses are formulated for conducting the study. The hypotheses are examined and analyzed regarding the data collected through a questionnaire. Based on the study's major findings, the hypotheses are tested and presented below.

H1: Quality of Work Life (QWL) in the libraries attached to the universities are the same.

Walton's eight-point dimension on QWL was modified to conduct this study. The analysis reveals that the Quality of Work Life (QWL) related to libraries attached to the universities is different. It is revealed through findings. Hence, the null hypothesis formulated for the study was rejected, and an alternative hypothesis that the Quality of Work Life attached to the libraries of the university is different.

H2: There is a significant association in Quality of Work Life (QWL) among LIS professionals in the universities of Kerala.

The study revealed a significant difference in the QWL of the LIS professionals working in the universities in Kerala. It is revealed through the findings based on the perception of LIS professionals. Hence, the result of the study rejected the null hypothesis formulated. The alternative hypothesis formulated from the study is that there is a significant difference in the QWL among LIS professionals working in the universities of Kerala.

**Suggestion**

Given the above study status, the following suggestions are made to improve university libraries' Quality of Work Life (QWL).

- Improve employees' perception of the various dimensions of Quality of Work Life in libraries and information centers. It will improve the working conditions of libraries.
- Improving QWL in libraries will motivate employees and enhance their organizational performance and morale.
- Comprehensive training for new entrants. It helps to be fully acquainted with the organization.
- Effective counselling for employees. It helps professionals enhance the realities of their work and social lives.
- The works of the professionals should be scientifically determined after considering the working conditions of the organization.
- The strength of the professionals in the libraries should be increased in proportion to the workload.
- Unity and a positive approach among the workforce in the university should be encouraged to satisfy the information requirement of the user community.

- Authorities should be sympathetic to the genuine concern of the professionals working in the libraries. It will develop a sense of involvement among the LIS professionals.
- The professionals should realize their changing role and contribute accordingly to the organization due to digital transformation and disruptive technologies.
- The policies of the organization should be framed considering the long-term consequences.
- Professional library associations have a prominent role in improving QWL through Continuous professional development (CPD) programs.
- The availability of human resources in the university libraries should be reviewed, and action may be taken to fill up the vacancies of LIS professionals for the effective functioning of the university libraries.
- The LIS professionals should frequently conduct orientation and user awareness programmes among the user fraternity to the maximum utilization of available and accessible online resources in the libraries.
- New entrants of the LIS professionals should have opportunities for advancement and successful careers in the universities.

**Suggested Areas of Further Research**

The following areas are identified and suggested for undertaking similar studies.

1. One may take up several similar libraries and make a comparative study and continuous research analysis in other libraries, such as academic libraries, special libraries, and public libraries. It helps to better understand the Quality of Work Life of other libraries and information centers.
2. Quality of Work Life and the service quality of LIS professionals of the library and information centers can be analyzed.
3. Quality of Work Life is an essential aspect of an organization. A detailed investigation on the impacts of job motivation on library and information professionals can be studied.

## Conclusion

In this study, an attempt has been made to determine the Quality of Work Life (QWL)of Library and Information Science professionals working in the universities of Kerala. Based on this analysis, suggestions and further research areas have been put forward for improving QWL in university libraries. Quality of Work Life is related to the expectations of individuals and the different characteristics of a job. The study will help the authorities redesign the policies of service organizations like libraries. It will open a new window in the emerging teaching, learning, and research environment. The study will set a new benchmark and help all stakeholders of the universities enhance the quality of work life in the university libraries.

# Bibliography


Agrawal, Amee (2019). Management Perceptions of the Impact of Quality of Work Life on Organizational Performance. Prabandhan: Indian Journal of Management.12(8),32-45.

Ahmadian, R., Vafaeian, M., & Farshbaf, M. (2015). The Relationship between Quality of Working Life with Organizational Commitment and Employees Productivity of Physical Education Faculties in the Islamic Azad Universities of Tehran. International Journal of Life Sciences, 9(4), 49-53.

Anandh, K. S., & Gunasekaran, K. (2016). Constructing a Model to Examine the Influence of Quality of Work-Life on Work-Life Balance-Discernment of Civil Engineers from Construction Industry in Chennai. Indian Journal of Science and Technology, 9(40)

Annual Report 2019 (2020). University of Calicut.

Annual Report 2019 (2020). University of Kerala. Thiruvananthapuram: University of Kerala.

Anyaoku, Ebele N. (2016) Demographic Determinants of Quality of Work Life of Librarians Working in Nigeria. International Journal of Advanced Library and Information Science, 4(1) :312-323.

Bhende, Pravin et al. (2020). Quality of Work Life and Work–Life Balance. Journal of Human Values. 26(3), 256–265.

Bishop, B. W., Orehek, A. M., & Collier, H. R. (2021). Job Analyses of Earth Science Data Librarians and Data Managers. Bulletin of the American Meteorological Society, 102(4), 1-22.

Darmawansyah, Rompu., Wahyu Atjo., & Abadi M.Y. (2018). The effect of Occupational Stress, Quality of Work life and Organizational Climate on Officials' Work Satisfaction of Regional Public Hospital of Undata Palu, Indian Journal of Public Health Research & Development, 9 (3),309-313doi: 10.5958/0976-5506.2018.00228.0

Darmawansyah, Rompu., Wahyu Atjo., & Abadi Muh. Yusri. (2018). The effect of occupational stress, Quality of Work life and Organizational Climate on Officials' Work Satisfaction of Regional Public Hospital of Undata Palu. Indian Journal of Public Health Research & Development,9(3),309-313. doi:10.5958/0976-5506.2018.00228.0

Davis, L. & Cherns, A. (Eds.). (1975). The quality of working life. New York: Free Press.

Davis, L. E. (1983). Design of new organizations. In Harvey, K., Hansvan, B. (eds.). The Quality of Working Life and the 1980s. Praeger Publishers, 65–86.

Economic Review,2020(2021). Thiruvananthapuram: Kerala State Planning Board. Vol. I.

Economic Review,2020(2021). Thiruvananthapuram: Kerala State Planning Board. Vol. I.

Economic Review 2020. (2021). Vol. I. Thiruvananthapuram: Kerala State Planning Board. Retrieved from https://spb.kerala.gov.in/sites/default/files/2021-01/English-Vol-1_0.pdf

Fakhri, Mahendra et.al. (2020). Characteristics of Quality of Work Life on Employees at Consultant Company in Indonesia. Journal of Asian Finance, Economics and Business, 7 (11), 1105–1111.

Garg, Babita. & Surya Nath Singh. (2018). Measuring the Relationship between Quality Work Life and Organizational Environment: A Case Study of Library and Information Science

Professionals of Haryana State. IOSR Journal of Humanities and Social Science (IOSR-JHSS), 23(10) :14-19.

Ghasemi et al., (2021). Quality of work life among surgeons and its association with musculoskeletal complaints. Cogent Psychology., 8. pp.1-16,8: 1880256

Ghosh, Suresh C (2007). History of Education in India. Jaipur: Rawat Publication, p. 452.

Gupta, R., & Kumar, S. (2020). Students Perception for Adopting Marketing Tools in University Libraries on the basis of their Backgrounds. A Study of North Western India. DESIDOC Journal of Library and Information Technology, 40(6), 339-344.

Harrison, T.M., (1985). Communication and Participative Decision Making: An exploratory study. Personnel Psychology. (1), 97-116.

Hattangadi, Vidya. (2015). Theory X & Theory Y. International Journal of Recent Research Aspects. 2: 20–21.

Herzberg, F. (1968). One more time: How do you motivate employees. Boston: Harvard Business Review. pp. 46–57.

Herzberg, Frederick., Mausner, Bernard & Snyderman, Barbara B. (1959). The Motivation to Work (2nd ed.). New York: John Wiley.

https://covas-opac.l2c2.co.in/cgi-bin

https://hgcl.kannuruniversity.ac.in/Pages/index.html

https://inflibnet.ac.in

https://library.ssus.ac.in/index.php

https://libraryscience.uoc.ac.in/

https://mguniversity.ac.in

https:// old.mu.ac.in.

https://uoc.ac.in

https://www.culibrary.ac.in

https://www.cusat.ac.in.

https://www.kannuruniversity.ac.in/en/academics/

https://www.kvasu.ac.in/

https://www.unom.ac.in/

Hulin, Charles L., & Blood, Milton. R. (1968). Job enlargement, individual differences, and worker responses. Psychological Bulletin, 69(1), 41–55.

India (2021). Report of the Comptroller and Auditor General of India on Economic Sector for the year ended March 2019. Government of Kerala Report No. 4 of the year 2021.

India: (2020). National Education Policy 2020. New Delhi: Ministry of Human Resource Development.p.52

Jahani, M. A., Ghahramanmahmoudi, Yaminfirooz, M., & Shahrdami, Y. (2017). Structural Model of the Relationship between Perceived Organisational Support and Quality of Working Life of Employees of the Selected Hospital of NorthofIran. Ambient Science, 4(1),23-27.

Johannsen, H. and Page, G. T. (1980). International Dictionary of Management. New Delhi: Vision Books; p. 277. https://doi.org/10.1007/978-1-349-05769-6

Kamboj, S., Ramnath, & Singh, S. (2016). Emotional Correlates of Quality of Work Life: A Study on Call Centre Employees. Indian Journal of Health and Wellbeing, 7(9), 940-942.

Kaula, P.N. (Ed.) (1958). Library Movement in India. Delhi: Delhi Library Association, 1958.p.58.

Kerala (2021). Report of the 11th Pay Revision Commission Government of Kerala., Part III.

Koohbanani, Hamta Kazemi. & Zarei, Atefeh. (2019) Measuring the librarians' quality of working life in the Iranian public libraries, Library Management,40(8,9),532-542. © Emerald Publishing Limited, 0143-5124. DOI 10.1108/LM-11-2018-0089

Lawler, E. E. (1975). Measuring the psychological quality of working life: The why and how of it. The quality of working life, 1, 123-133.

Lawler, E.E. (2005). Strategies for improving the quality of work life, American psychologist, 37, 486-493.

Maslow, A (1954). Motivation and personality. New York, NY: Harper.

Maslow, A. H. (1943). A theory of human motivation. Psychological Review, 50 (4): 370–396.

May, B., Lau, R., & Johnson, S. K. (1999). A longitudinal study of quality of work life and business performance. South Dakota Business Review, 58, 1-7.

Mothukur, Anjaiah. & Ramesh, Elagandula. (2018). Quality of Work Life of Library Professionals in Selected Autonomous Engineering College Libraries of Hyderabad. International Journal of Library and Information Studies, 8(2): 339-345.

Naik, J P (1982). The Education Commission and After. New Delhi: Allied Publishers Private Limited.p.1

Nandy, Subodh Gopal. (1985). Job satisfaction of the library professionals. Herald of Library Science.24(4), 295-300.

Nanjundeswara swamy, T S & Swamy, D R. (2013). Quality of Work life of Employees in Private Technical Institutions. International Journal for Quality Research 7(3) 3–14.

Naskar, D., Hasan, N., & Das, A. K. (2021). Pattern of Social Media Engagements by the Learners of a Library and Information Science Mooc Course: An Analytical Study. Annals of Library and Information Studies, 68(1), 56-66.

Neelameghan, A. (2007). Library is a growing organism – S R Ranganathan. DLIBCOM, 2 (8-9), 2-3.

Netto C S (2017). Quality of Work Life of women employees in private sector organizations in Ernakulam District of Kerala. Prabandhan Journal of Management.11(11),51-62.

Omole, Victoria Imabong (2020). Relationship among Job Satisfaction, Leadership Styles and Career Indecision of Library Personnel in Selected Private Universities in South-West, Nigeria. International Journal of Library Science, 9(2): 40-50

Prytherch. Ray. (19950. University Library. In Harrod's Librarians' Glossary (8th ed., p.663). England: Gower.

Rachel, Rabeka., & B, Mini Devi, B (2019). Job Satisfaction of Library Professionals of University of Calicut: An Analytical Study. Indian Journal of Information Sources and Services, 9(1), 132-135.

Raja E A., Prabhu, N R V., & RAO, P K (2010). Managerial behaviour and effectiveness (P.116). New Delhi: Excel Books.

Ramawickrama, J; Opatha, H.H.D.N.P & Pushpakumari, M D (2018). Quality of Work Life and Job Performance: Empirical Evidence from Station Masters Working at Sri Lanka Railways. Sri Lankan Journal of Human Resource Management, 8 (1) ,1-21.

Ranganathan S R. (1957). The Five Laws of Library Science, 2nd ed. Bombay: Asia Publishing House.

Rao, Mahendra Singh (2020). Quality of Work Life (QWL) Motivates Library Professionals Highly to Learn Further and Develop Skills. Library Philosophy and Practice (e-journal). 4618. https://digitalcommons.unl.edu/libphilprac/4618

Rao, Mahendra Singh. & Verma, L. N. (2019). Quality of Work Life among Library Professionals: A Study of Determinant Factors. SRELS Journal of Information Management, 56(2):95-99.

Reddy, Pratapa B. (2010). Quality of work life of professionals in public libraries in Andhra Pradesh - A study. PEARL - A Journal of Library and Information Science,4(2), 95-103.

Reghagwa, I. C., & Ijiekhuamhen, O. P. (2020). Influence of Information Communication Technology Competencies on the Task Performance of Librarians in University Libraries in South- South, Nigeria. Library Progress, 40(2), 204-214.

Seema, Arif., & Maryam, Ilyas. (2013). Quality of work-life model for teachers of private universities in Pakistan. Quality Assurance in Education, 21(3), 282 – 298. http://dx.doi.org/10.1108/QAE-Feb-2012-0006.

Shanmugam, S &. Bharathi, Subburethina P (2017). A Study on the Influence of Socio-Economic Factors on Quality of Work Life Perceptions among Self-Financing Engineering College Teachers in Tiruchirappalli District. Shanlax International Journal of Management,5(1),147-158.

Sharma, Narendra Nath. (2016). Job Analysis of Library Professionals. In Science and Technology Institutes of Assam:An Evaluative Study (Ph. D. Thesis), Gauhati University, Guwahati. Retrieved from http://hdl.handle.net/10603/144701

Shyni, K.G. & Muhammed Haneefa, K. (2016). Influence of Job Satisfaction on Quality of Work Life of Library Professionals in University Libraries in Kerala. KELPRO BULLETIN, 20 (2): 64-71.

Suresh, K (2019). Empirical study on the Quality of Work Life of Nurses in Private Hospitals. Shalax International Journal of Management.7(1),92-102.

Swayam, "Emerging Trends & Technologies in Library & Information Services (ETTLIS)," https://swayam.gov.in/nd2_arp19_ap78/preview.

Tamsah, H et.al (2020). Training, Knowledge Sharing, and Quality of Work-Life on Civil Servants Performance in Indonesia. Journal of Ethnic and Cultural Studies, 7(3), 163-176.

Tiwari, N., & Singh, K. P. (2018). Human Resource Development in Libraries: A Literature Review. Library Herald, 56(2), 241-254.

UGC (2018). UGC Regulation 2018 notified in the Gazette of India, Part III, Section 4, July 18, 2018. Available at https://www.ugc.ac.in/ugc_regulations.aspx

Varadaraj, S., & Parthiban, V. 2009. Impact of Demographic Variables on Quality of Work Life of Supervisors. SMART Journal of Business Management Studies, 5(2),47-52.

Walton, R. E. (1975). Criteria for Quality of Working Life. In: Davis, L.E., Cherns, A.B. and Associates (Eds.) The Quality of Working Life, The Free Press; 1975. P.91-104.

Walton, R.E. (1975). Criteria for Quality of Working Life. In L.E. Davis, A.B. Cherns and Associates (Eds.), The Quality of Working, New York: The Free Press, Life, Vol. 1, pp. 91- 104

Walton, Richard. (1973), Quality of work life: What is it? Sloan Management Review. 15(1), 11-21.

# Index